POLICEMAN'S EVIDENCE

POLICEMAN'S EVIDENCE

by

Rupert Penny

RAMBLE HOUSE

ISBN 13: 978-1-60543-041-6

ISBN 10: 1-60543-041-2

Cover Art: Gavin L. O'Keefe
Preparation: Fender Tucker

POLICEMAN'S EVIDENCE

FIRST PART

I

CHIEF-INSPECTOR EDWARD BEALE—doubtless—turned the envelope over idly in his hand before opening it. He was, almost certainly, wondering what could have induced myself, his friend Anthony Purdon, to write at all in the first place, and at such obviously great length in the second. It may have crossed his mind for a moment that perhaps when he looked inside the envelope he would find only a batch of newspaper cuttings, or a couple of ties to brighten the landscape when he was off duty; but he would have been wrong. What he had just received from me was twenty-four closely written pages.

I suspect that before he began to read my letter he poked the fire into a blaze and drew his arm-chair nearer. The first days of January were bitterly cold, and the nights worse, and the feet of policemen are notoriously delicate, even if they have long ago ceased pacing the dark lonely streets and peering dutifully at all that moves nocturnally. He would also have been smoking a pipe: that is a safe bet.

> Mauberley Grange,
> Gloucestershire.
> January 10th, 1938.

Dear Ted,

Forgive the shock—I'm certain it'll be one. Forgive also absence of weather reports, health enquiries, and so on. There is a great deal to write about that's more interesting. I told you at Christmas that I was going into the country for a week or two, but not why or wherefore. I had half promised not to, but now have full permission.

Did you ever hear of Major Francis Adair, D.S.O.? He was in the Intelligence Service during the War, and for quite a while afterwards, and was one of their crack experts on codes and ciphers and such-like inventions of Beelzebub. Later he went abroad, 1925-33, poking his nose into a number of pies: South America, China, Egypt. Then he

came home after a bad bout of fever, and settled—surprisingly—in one of the genteeler parts of Surrey. He's a
queer old boy: perhaps not really so old, say about sixty,
but definitely queer. He's got one of those beaky faces with
a distinguished thatch of white hair, and his voice is very
parade-ground and H. M. Bateman. I'm sure he'd have only
had to bark twice to be a general, but he never seems to
have bothered. At all events, he's queer—sorry for the
repetition. For one thing, he's mean—really mean about
some things, like housekeeping money and tobacco. I've
kept him in cigarettes for the past week and been expected
to, what's more. Yet he isn't wholly mean, because over big
things he can be quite casual and careless. That comes
later, though.

To tell the story right way up: Adair is a widower with one
real daughter and one adopted one, former called Tilly and
latter Lina Hipple, former a pasty spectacled wishy-washy
creature who smells of carbolic soap and invisible curates
and dresses like a jumble sale, latter very soignee and
scented, an orchid drenched in patchouli. The 'hip' of Hipple
is much in evidence; likewise neighbouring features. She
has long dark eyelashes to match her dark hair, and she
droops them at you—yea, even at me, forty all but and ugly
as Anubis. Still, I am male, so I suppose I do to practise on.

Adair also has one secretary, a smart fellah by the name
of Hinkson, all knife-edged trousers and cuffs and brillian-
tine; he'd never be my choice, but apparently he knows
how to make himself useful. Then there's a butler, Andrews,
rather depressing on the whole, though with his point of
interest—later; and a chauffeur-gardener, Judd, a surly-
looking brute but a wizard at the wheel. He has a wife, and
I've seen her, but I can't remember what she looks
like—very faded and meek, I fancy. As well there is a couple
of hundredweight or so of cook, of no particular merit as
such, and one Buck, a bodyguard.

Yes, the address is Gloucestershire, not gangsterland,
but that's where I should think he originally came from. His
real name is Kass, so he says, but nobody ever calls him
anything but Buck. He's small, thick, nearly bald, has
steady blue-black eyes, hardly any finger-nails, and bad
teeth. I wouldn't very much like to quarrel with him unless
I were allowed the first three shots, but he's pleasant
enough, though he and the shiny Hinkson don't get on too

well. As for the reason for his presence, that will come out in due course.

To complete the party there's Roger Montague, another queer bird. He must be considerably older than Adair, and it seems they've known one another since the end of last century, but you can scarcely imagine a bigger contrast. One's big, bluff, a civilized bully—that about sums up Adair, for all he's my host; and the other's small and thin and delicate, with semi-transparent flesh and a bored cynical smile. He doesn't say much, but it's often to the point, and mostly tart in flavour. Nor is he at all active—partly because of his poor digestion, I think. He lives chiefly on dry toast and barley-water, with an occasional blow-out of weak tea and steamed fish: enough to make any one bored and cynical, I agree.

And now, what in the world are we all doing here? Well, that should probably have come first, but my letters never work that way.

Not to conceal the facts of the matter any longer, we're all taking part in a treasure hunt.

Real treasure, we hope, and certainly a real hunt.

If you can persuade yourself to read on, I'll explain how it all comes about. I said that Adair was a code expert: he reads Bacon's bi-lateral cipher as easily as you or I might read Areopagitica or Sordello. Likewise with the various shorthands—Pitman, Dutton, Sloan-Duployan, Gregg, Gabelsberger, Newton-Rapid, and all the rest of them. Well, about eighteen months ago he was nosing round at a country auction sale when he came across an old leather-bound volume containing about three hundred pages of a kind (of shorthand) he hadn't met before. He bought the thing for ten bob, as an ordinary person might buy a book of crossword puzzles, to provide amusement for the next rainy day. However, even for Adair it wasn't as simple as that. In fact, it took him quite a time to get started at all, and once he'd managed that it wasn't plain sailing, because the author hadn't been very particular about keeping to his own rules: needless to say, it was a private shorthand.

The translation, which I've seen, isn't of much interest except for one part of about twenty pages, which relates the history of the Mauberley family. Its last survivor, Jasper, was born in 1634, just after this house was built, and died in 1706, and all his life he was afflicted with a hump on his back and an impediment in his speech. These handicaps seem to have soured his nature, because from early manhood to the end of his days he lived here almost alone, never going anywhere, never having visitors, and of course never marrying. In short, he emerges from the story as a cantankerous crippled miser.

When he passed on, however, there wasn't a trace of his wealth to be found; nor, naturally, many people who minded much—no surviving relatives, as I've said. Some half-hearted sort of a search was made by inquisitive neighbours, but nothing resulted, and it was eventually agreed that he must after all have been poor instead of mean. The author of the leather-bound shorthand diary disagrees, though. He, by the way, was a Boon: Thomas Stanway Boon, Esquire, 1741-1788, also of Mauberley Grange. He produces very little in the way of fact to support his opinion, but twaddles a bit about having seen a mysterious hunchbacked figure in his room one night wearing an expression 'of the utmost secrecy and ill-natured satisfaction'. The visitation left him convinced that somewhere in this rambling old place was Jasper Mauberley's treasure, and he made frantic but unsuccessful attempts to locate it. In the end he gave the job up as useless, merely relating the facts in his home-made shorthand, repeating his opinions, and wishing the heartiest good luck to anybody that cared to have a go in the future.

'I am persuaded,' he says 'that the clue to the whole matter should be found in the panelled room at the near end of the Long Corridor upon the ground floor. There are still living, though old now and not wholly in their wits, it may well be, those who declare that in this room alone did Jasper Mauberley pass the twenty years before his death, bed-fast towards the end. I incline to believe that the treasure, if any exist, will not be discovered there, nevertheless, but in another place: but that the old man, confined within the house and unable to visit his secret hoard, may well have concealed somewhere about his room that which will point the way, if there should be any to find it, or, having found, to interpret the same.'

Now, I think I told you earlier—can't be bothered to make sure—that Adair's got plenty of money. He's also got a strain of avarice in him above the usual—I know I mentioned that. Yet it isn't the niggling avarice of the man who's afraid to spend twopence on the chance of winning a shilling, because he might lose. Adair would spend his twopence, and then devote his whole energies to making sure he did win. So with the present case: the idea of a treasure hidden somewhere, his for the finding, appealed to him.

Useless to suggest—I did try—that maybe there never was one, or that anyway there's been time for a dozen people to get in before him. He wanted the treasure, therefore it was obtainable. It was obtainable—therefore it was merely a matter of time before he would be reckoning its value. Mule's logic, perhaps, but the sort that won't be daunted. He at once made enquiries about this Mauberley Grange, found that it was in the market, and got a fortnight's option on it at a price of £6000. The last Boon died in 1933, and the place has been vacant ever since. It stands in its own ground of three hundred acres on the Hereford-Gloucester border, can't be seen from any road because of trees, and might easily go unsuspected by a stranger, and the cultivated part is in a simply shocking state.

Even twenty years ago the house must have been a fine place, though considerably built on from its original design, and not always to match, but the last of the Boons seems to have let it go completely, and four years' vacancy haven't helped. Parts of the roof are in ruins, the west wing was half burnt down in 1934, the drains are primitive, and a lot of the interior woodwork has been eaten away. The open parts near the house are shockingly neglected, too: orchards overgrown to glory, lawns like a wilderness, the drive little better than a cart-track. I can't decide whether £6000 was cheap or dear. There are forty-two rooms in the place, and a hundred acres and more of so-so timber, and the prospect of a treasure: what do you think?

Well, next I'd better explain about Buck the bodyguard, and the reason for his presence. Half way through the translation of the diary—in fact, about half way through the tale of Jasper Mauberley—Adair had occasion to sack his secretary for theft. That was roughly eight months ago, the chap's name was Warner, and in his place Hinkson was engaged, chiefly on account of his knowledge of ciphers: not by any manner of means as extensive as Adair's own, but far superior to most people's. Then, a couple of months later, during temporary absence of owner and household, there was a burglary at his Surrey home. Nothing seemed to have been taken, and it was Hinkson who suggested what might have happened. He was positive that somebody had handled the diary—carelessly not locked up—during their absence, and his idea was that the dismissed Warner had sneaked back, photographed the rest of the stuff, and might now be hard at work solving it. He'd be pretty familiar with the key to the shorthand, of course, having worked with Adair.

The old boy didn't take the suggestion very seriously at the time, but at the end of the same month, September, when negotiations for the purchase of the Grange were on the point of completion, the agents tried to make a hitch. Apparently they'd received an urgent telephone call from a man calling himself Duffy, offering £7000 hard cash for an immediate deal. Fortunately Adair was covered by his option, which he immediately exercised, and Duffy never materialized, so nobody knew who he really was, or even if. Hinkson declared him to be Warner, and persuaded Adair, but I have a sort of suspicion myself that he was that not too rare creature, a house-agent's myth.

However, as I said, Hinkson shouted, 'Warner, by cripes!', the Major duly echoed him—and the result was Buck. Where he dug the little chap up from I can't say: I don't like to ask Adair, and Buck himself just grinned when I tried to pump him, and murmured something about a game of poker on a ship. He carries a gun—for which Adair somehow wangled him a certificate, so don't start worrying—and he can use it, too. He demolished a slug on a gate at about fifteen yards very neatly. I wouldn't bet he'd never been on the wrong side of the law, and as a matter of fact I've tried my damndest to get his fingerprints, in case you've got a copy in the family album, but without success. Can it be that the little wretch suspects my innocent ruses? Could anybody?

Well, there isn't much more m the way of facts. Here we all are, searching rabidly for the clue to the treasure, but so far empty-handed. If you've got nothing to do, I could probably get you an invitation to join us. To be honest, I rather wish you were here you might be able to see farther than I can, or else tell me there isn't anything to see. I'm

the least bit worried, Ted: not for myself, but by the atmosphere. There's something wrong with it, and I can't say
what, and it annoys me.

Lina and Hinkson, now: am I only imagining, or is there
something between them? Again, Buck and Hinkson: they
don't like one another—why? Item, Tilly: Montague's the
only person in the house barring self, who do my humble
best, who gives her a civil word. Adair at times is fiendishly
rude to her: you can see him preferring Lina almost audibly.
And anyway, who the devil is Lina? 'Daddy was a very dear
friend of the Major.' 'Her father was a spy, I think.' (From
Tilly, that: but of course Adair himself was one too, only on
our side, which takes away the nasty taste.) 'Hipple? My
dear boy, how should I know? Obviously a personage, to
judge by results.' (That's Montague: but surely he ought to
know—isn't he one of Adair's oldest cronies? Or isn't he?)

Item, Judd the chauffeur: why should he have one of
Lina's handkerchiefs in his pocket? That he did, I know; it
dropped out, and her perfume is obtrusive. I returned it to
her, and she didn't shift a hair: yet it was a green silk
specimen that she couldn't possibly have forgotten—far too
vivid a green. Judd had found it, and meant to return it at
the first opportunity? All right: why was it in an opened
post-marked envelope addressed to her? Item the last:
why does Andrews the butler write left-handed, although
he uses his other one for everything else? Is that natural, or
significant?

But enough of these pointless speculations. As you aren't
here, you can't possibly tell me the answers. I must turn in
now, because we've worked hard. The finder of the clue is
to receive a tenth part of the treasure, and you can see our
several faces glistening with greed when we think nobody's
looking. Except Montague's—he doesn't believe in treas-
ures, or says he doesn't. All for now. Don't get put on a
murder case, or I shan't know which way to turn.

Yours ever, Tony.

II

You may perhaps think that was a silly rambling sort of
letter to write, and I have no excuses. Doubtless it was, but
I print it to show my honesty: it is almost my only claim to
your consideration. What I saw for myself I will faithfully

retail in my share of this book, and what I failed to see you will have to do without, or supply for yourselves between the lines.

Well, I think the initial facts are pretty clear. There were nine of us actually inhabiting Mauberley Grange from New Year's Day onwards: Major Francis Adair, Tilly his daughter, Lina his adopted daughter, Roger Montague his old friend, Hinkson his secretary, Buck his bodyguard, Andrews his butler, his anonymous cook, and myself, Anthony Purdon, assistant editor of *The Stockbroker* and club acquaintance of Adair. In addition, Judd the chauffeur and his pallid wife lived at the tumbledown lodge at the bottom of the drive. Lest the shortage of servants cause surprise, I had better say at this point that two girls came in from the nearest village daily, Sundays excepted. They were pleasantly slow-spoken and civil, but have no part in this tale.

Although Mauberley Grange had been vacant for over three years, so great was Adair's anxiety to begin looking for the clue to Jasper Mauberley's treasure that he wouldn't waste time—or money, perhaps—in having it properly put to rights. It received only those attentions which were absolutely essential. The chimneys were swept and re-paired where necessary, the water-supply was tested and a new pump installed together with various storage tanks, and all broken windows were re-glazed. These things, with certain other requisite alterations, and about £250 worth of second-hand furniture, had to suffice to make the huge old place habitable.

Of the forty-two rooms we furnished only fifteen: it would clearly be far simpler to explore what was empty than what wasn't. Nine of the fifteen were bedrooms, and the rest apportioned as follows: one study, for Adair's own use; one dining-room; one kitchen fitted with oil cookers in addition to the built-in range; one living-room for Buck, Andrews, and the cook; one for the use of the two girls; one for Montague, Hinkson, and myself; and one designed, as someone said, for the politer commingling of the sexes. Luckily Adair wasn't mean about fuel—though possibly that was because Hinkson had seen to its ordering. We could have fires in our bedrooms day and night if we wanted them, and so could always escape unwelcome company.

Now, I don't want to dwell on myself unduly, but I'd like to make clear my relations with the other members of the party. Adair himself I can't honestly say I ever cared for much—most of the liking was on his side. I must admit, though, that as far as I could see his faults were chiefly on the surface, and you were all right with him provided he approved of you. His standards of approval may have been based on some coherent foundation, but if so I never discovered it, and I fancy that like the majority of people he was guided almost entirely by a blend of instinct and caprice.

(To branch off for a moment from my own case, and to consider his: he tolerated Buck and the rest of the servants, he was friendly enough in his peculiarly gruff way with Montague and Hinkson and myself, and he was clearly enamoured of Lina. This, in my opinion, was an exhibition of bad taste, and one I found it hard to account for. Externally she was admittedly attractive, and a casual observer might easily have taken her for his mistress rather than his adopted daughter; yet I am perfectly certain that there was no such relationship. He treated her more as a particularly adorable kitten whom he delighted to spoil, and she of course played up to him.

With his own daughter Tilly, on the other hand, he was on no sort of good terms whatever. Poor Tilly! I was really very sorry for her. Plain to the point of ugliness she might be, and devastatingly earnest and humourless, but even so he was not on that account excused for treating her the way he did—as if she were an especially unwelcome guest whom he was endeavouring to get rid of by every method except the direct one of telling her to get out. Sometimes he went so far as to ridicule her openly when she failed to grasp some point he was making: though Lina could behave like a fluffy-minded imbecile, and he would only smile at her benevolently. Tilly, as a matter of fact, was no fool over things in which she was interested. She had an honours degree for classics, and could discuss politics intelligently without party bias, but she knew very little about codes and shorthands, and her attempts to pretend interest in her father's pet subjects lacked conviction.

She deserves mention in one other respect: rarely could any female have displayed less clothes-sense, I thought, after living in the same house with her for a week. It was easy to imagine that all her various garments were stuffed piecemeal in a trunk, and that each morning when she rose she would shut her eyes, plunge her hands inside, and draw out at random enough of everything. One day we break-fasted with a blue tweed skirt, a tartan blouse, a pair of evening slippers, and a linen smock with two buttons missing; the next with a flowered silk frock, brogue shoes, a navy-blue blazer edged with faded gilt braid, and woollen stockings, odd as likely as not. In short, the girl invariably looked a freak: yet she was good-hearted, and interesting to talk to when she forgot to be nervous or instructive, and at times it was embarrassing to see how people cold-shouldered her. With Lina, of course, she had nothing in common but age—they were both about twenty-two. Hinkson certainly smiled incessantly when he was speaking to her, but he did that with everyone, and it was noticeable that he always began to edge away after a couple of min-utes. Montague was somewhat more considerate, but I really believe that apart from myself—I hope—only An-drews made no distinction between her and the rest, ac-cording her the same silent deference and attention. Have I said that he was a silent man? He walked more nearly like a cat than any human being I have ever known, and yet without giving the impression of furtiveness. Only his eyes did that, even when regarding you most steadily.)

I agree: it was about time I closed that parenthesis and returned to my original subject. I've said that with Tilly I tried to be reasonably polite and friendly; but with Lina I can't pretend I troubled much. I felt that where so many men paid court—Adair, Montague, Hinkson, Judd, even Buck—I might be excused. It may have been deplorably bad manners on my part, but I just wouldn't bring myself to be more than civil. She might ogle and sway at and vamp me to her heart's content; I determined to ignore the performance, and as far as possible the performer. Anyway, I detest females who can't be in a room for five minutes without making it smell like the scent department at Har-rod's. I also loathe and execrate magenta fingernails, in my view an infallible sign of ill breeding. Furthermore, I distrust anybody who maliciously teases animals, and she received a very black mark against her from me the morning I looked out from my bedroom window to see her strand the cook's black cat in the middle of the drive, a paper bag firmly fastened over its head with a rubber band, and then indolently watch its frantic backward progress. Her comment, in answer to my bluntly expressed disapproval, was typical.

"All right, Sir Galahad, keep your blood-pressure down—it's only a cat, you know."

Of course I said the obvious thing about the duty for compassion towards one's fellow creatures, and after that she swayed a little less in my direction.

With Montague I got on better. He was at times amusing in his slightly bitter way, and although in a perpetually poor state of health he could talk of something other than his ailments.

Hinkson I could never be sure about. One day I would almost think I liked him, and the next I would decide all but definitely that I didn't. He was always perfectly polite and amiable—just the least bit too much so, I felt occasionally: but in repose there was a touch of the wolf in his face, in the way his lips curved fully and in the slant of his tapering nose. I don't usually reckon a man by his appearance, but nor do I entirely discredit it, and I didn't take to his.

Perhaps the person out of the whole collection with whom I got on best was Buck. Judd struck me as too much the bruiser—his forehead low, his ears misshapen, his gash of a mouth unnatural without its chewed and drooping cigarette. Andrews seemed rather a nonentity, a rubber-soled wraith who whispered instead of speaking and whose servile attitude worried me: genuine servility is even harder to bear with than false, and his appeared genuine enough. In a way he was the male counterpart of Mrs. Judd, that wistful washed-out wretch. But in Buck I found plenty to interest me.

He was a good companion provided you didn't try to probe too deeply into the true nature of his character and opinions, and he had that merry twinkle in his eye which is either granted a man at birth or forever withheld, since it can't be cultivated. In brief, he gave me the impression of being more vital than any of them save perhaps Adair, and I'm not yet so anaemic myself that I find vitality depressing.

Don't let me put anybody on the wrong track about my feelings towards Buck, though. For instance, I'd no more have trusted him with a confidence that mattered than I would have kissed him. He inspired me with no conviction of honesty—rather the reverse: yet somehow with him it didn't matter. He might easily be a rogue: but what in the world else do you expect a gunman to be? Adair had hired him, as both told me at various times, because he was tough and alert, and had some experience of shouldering other people's troubles, of the kind that might be expected if Warner were engaged in a simultaneous and illicit treasure hunt. Also, Adair had said, he believed that Buck was straightforward with his employers. He might go over to the enemy if he thought it would pay him better, but he'd give notice of his intentions first.

It had been strictly impressed on him that his present duties amounted to no more than the vigilant use of eye and ear. If he should come upon a stranger in the grounds he was to deal with the matter in the English fashion: ask questions first, and shoot, if at all, as the last possible alternative to himself being shot. When he found time to sleep I couldn't imagine, for he was active both by day and by night: but the day-time he treated more as play than as work, and it was not till it grew dark that his manner became business-like.

Then he would put on his old raincoat and slouch hat, and depart on his rounds, a gun in one pocket and a bag of acid drops in the other. He said they compensated for not being able to smoke.

III

We settled in on New Year's Day, a Saturday, but it wasn't till the Monday that we really began to search for the clue. Adair took sole charge—a majestic figure in worn grey tweeds, smoking a squat pipe or my cigarettes, and habitually shouting to anyone more than two feet away.

It was only natural, in view of Thomas Stanway Boon's opinion as expressed in his diary, that we examined first the panelled room at the near end of the Long Corridor. It was the only one in the house with no ordinary wall space, and had been purposely left unfurnished, and though it wasn't small—32' by 26' by 12' high—six of us would obviously get in one another's way. For this reason the girls were soon sent off to explore the three other vacant rooms on the same corridor: the fourth Adair was using as a study. They went reluctantly, but they went.

"Now," said the Major, "we'll do this business system-atically. Tony, you go round tapping the three top rows of panels—you're the tallest. Here's a walking-stick and some chalk—put a mark on any that sound queer.

"Hinkson, you'll take the lower three rows, and do the same. Roger, you and I'll share the floor between us. Any complaints?"

"Well," murmured Montague, "apart from the fact that I haven't the least idea what I'm supposed to be looking for, I really don't see what good all this tapping will be. Does a panel with a piece of paper behind it sound so very much different from one without?"

Adair's eyebrows bristled.

"How the blazes should I know?" he countered. "The only thing to do is to find out."

So each of us accordingly set about his appointed task: but neither Hinkson nor I had occasion to use chalk, and the other two didn't seem to be getting on much better.

The floor, incidentally, was of oak, fairly evenly laid and polished smooth by years of hard wear. The panels were also of oak, but blacker, and each measured two feet square. The top rows were plain, but those below had been carved with varying degrees of ornateness. According to Montague they were particularly good specimens, though both they and the floor were somewhat pitted with worm holes. There were two windows on the outside wall paned with small squares of thick yellowish glass, and the fire-place was huge and open. Personally I had more hope of success from the chimney than from anywhere, but the time for suggestions would doubtless come later.

As can be imagined, the room was dark, even with the windows wholly uncovered, and for this reason each of us had been supplied with an electric torch. In addition, we worked muffled up in coats and scarves, and Montague even went so far as to wear a rug in redskin fashion. We must have looked pretty stupid, if the truth were known.

At the end of half an hour the Major called a halt.

"Walls and floor change over," he directed, and we did so, but still with no result.

"How many rooms did somebody say there were?" asked Roger Montague presently in a dreamy voice. He was leaning against the door surveying Adair, energetically on tiptoe tapping, and his expression was chiefly one of tolerant amusement.

"Forty-two," answered Hinkson, with his ever-ready smile.

"Thanks. Then at a week each that makes about eleven months, and the gardens will still be in their present abandoned state, only rather more so. At least, if a mere year could make any noticeable difference, which on second thoughts I doubt. And there'll be the roof and the cellars and the foundations untouched, not to mention half a dozen secret passages, as likely as not."

"Drivel," said Adair shortly without looking round. "I told you yesterday there aren't any."

"I remember: but the nature of a secret passage is to be unsuspected and scoffed at till somebody comes out of or falls into it."

Normally the Major might have bridled at that, but today he was in a good humour, and merely grunted.

"Have it your own way," he said. "I got a couple of experts in seventeenth-century architecture down here to look the place over, and they assured me there weren't any."

"It is likewise the nature of experts to be right until they've been proved wrong," went on Montague, as dreamily. "And when that happens the event is dismissed as the kind of isolated accident that might happen to anybody. Don't you agree, Tony?"

"No, I can't say I do," I answered. "I object to the word 'isolated'."

"Excellent—so do I, really, between ourselves. Well, Francis,"—to Adair, "aren't you going to do the chimney next? Tony's simply longing for you to."

At this I turned towards him in some surprise.

"Now how on earth did you know that?" I asked.

"My dear chap, you've hardly looked at anything else since you entered the room. I've watched you watch an imaginary Jasper Mauberley climb into it, deposit his imaginary clue to his imaginary treasure, and laboriously climb down again."

"You ought to have been a detective," I said.

"Yes, perhaps," he agreed, a momentary smile on his lean lined face. "But then, everyone's a detective in his heart. Otherwise, what are we all here for?"

"Aren't you confusing detection with hunting, Mr. Montague?" queried Hinkson, in his politely amiable way. He hardly ever spoke without showing his even teeth, and I had begun to grow a little tired of their white monotony.

"Oh no," was the response: "merely equating them."

"Talk!" ejaculated Adair with sudden emphasis, as if he were referring to some incurable disease. "All these panels will have to come out, you know, but I suppose you'd better look up the chimney first. I think myself it's too obvious, though. Hinkson, you're the slimmest."

But the clue was not in the chimney: nor, to be brief about what took a long time to verify, behind any of the panels or under the flooring. We had everything movable out, in spite of Montague's acid comments about vandalism, but found nothing more interesting than an unsymmetrical bone button, fallen probably from some long-dead workman's clothes.

I'm not sure whether or not I ought to skip the details of the search, but I'm going to. In general, one day was very like another. We began soon after breakfast, about nine o'clock, knocked off for beer at 11.00, for lunch between 1.00 and 2.30, and stopped for the day as soon after 7.00 as we could manage without arousing Adair's disapproval. By the time we had finished with a room it looked like the scene of an earthquake or an air raid: we were really conscientious, if not of our own volition, then driven to it by the dauntless Major.

But, to give us our due, nobody shirked except Lina. For the first two days she was wildly enthusiastic; on the third she wilted a little; by the sixth her assistance was the slenderest apology for perfunctoriness, and she didn't hesitate to indicate that she considered the whole business a bore. In a way, of course, it was, but she alone declared her opinion openly, and I think we all felt that it was rather unfair of her to trade so on her adopted father's indulgence. I must admit, though, that she hardly looked right in the part of ardent seeker undiscouraged by lack of success. To regard her made you think more of heavily perfumed boudoirs and soft cushions and discreetly revealing wraps than of floors being dustily demolished, fireplaces poked and pried into, and panels removed to the discomfort of countless scuttling spiders.

The clue was at last discovered on the fifteenth day of the search, January 17th, a Monday; in a place where we agreed we ought to have looked earlier: only it is so easy to say that afterwards. It was lodged inside the left-hand beam forming the doorway to the paneled room, the join in the wood very cunningly concealed and only noticeable with the help of a magnifying glass. The finder was Adair himself, appropriately enough, though I think we were all disappointed that none of us would be able to claim a share of the treasure for services rendered. However, as you may suppose, we didn't all immediately pack our bags and depart.

The clue itself was a roll of yellowed parchment encased in a small wooden box, obviously hand-made and presumably by Jasper Mauberley: on the lid were carved the initials J.M., and the date 1695, and thus we had no doubt of its genuineness. But, as well, this box contained something else, a something quite unlooked for, which sent us into varying stages of exultation and delight: one unset ruby about the size of a pea.

"An earnest of the real treasure," said Roger Montague at once. "Has anybody noticed me scoffing these last ten days or so? If they have, let them kindly and permanently forget. Francis, I congratulate you."

And so did we all, clustering eagerly round the Major who made little attempt to hide his pleasure.

"I told you so!" he said. "I told you so! Hinkson, open up some champagne—we'll celebrate this. And I'll have mine out of a tankard—dammit, I found the thing, didn't I?"

He winked at no one in particular, slapped his thigh, and then carefully unrolled the sere ancient manuscript, several pages of it. Then his expression changed: he groaned, and we trembled in dismay.

"What's up, Nunky?" asked Lina. She was well to the front, her neat dark head peering over Adair's shoulder and her delicate cheek almost nestling against his.

"Look for yourself, Ladybird," he invited, in reply. (That was his incongruous nickname for her, though anybody less like one could scarcely be imagined. Ladybirds are homely innocent- seeming things: they do not pluck their eyelashes, nor do they wear blouses which gape provocatively at the breast.)

"But I am looking," she told him. "What's it all mean—all those queer lines and dots and things?"

"A month's work," he answered. "Two months'—maybe a year's. It's a private shorthand again,"—and once more he groaned.

"But hang it, you're an expert, aren't you?" I said. "What's a bit of shorthand to an expert?"

Before replying he painstakingly re-rolled the parchment sheets, eight or nine of them about 9" by 12", put them into their box, and slipped the whole into his pocket. The ruby remained in Lina's hand, I noticed.

"The trouble will be," he then said, "that there isn't enough of it."

"What do you mean?"—from Montague: all this time Tilly had remained almost silent in the background, and was standing now by the window polishing her spectacles. "Part of it missing?"

"No—at least, there's no reason to think so. But do for heaven's sake use your wits." He cleared his throat, extended his right index finger, and made it clear by his manner that my thoughtless question had provoked a lecture.

"I'm not forgetting that most of you know about as much of secret writings as I do of folk dancing," he began, "but doesn't it stand to reason that the shorter they are, the harder it's going to be to translate them? Any schoolboy nowadays can tell you that 'e' is the most frequently used letter in the English alphabet, but even that doesn't help in working out the most elementary cipher unless the thing's long enough for the rule to work. Exactly the same kind of trouble happens with shorthands—give me a cigarette, Tony, please. The principle involved—really, you ought to know at least this much—is the representation of sounds by signs: sounds, not letters. Well, they too can be graded in order of frequency for various languages: but what holds good for a whole book, say, needn't necessarily hold good for one paragraph, or even one chapter. If this"—tapping his pocket—"were in Pitman's, its translation would run to about seven or eight hundred words at most, and that including grammalogues. Nevertheless, call it eight hundred. Now, anybody who wanted to could easily write a message of that length without conforming to any known law of average sound frequency, and who's to say that

Jasper Mauberley wouldn't be up to a dodge like that? He was no fool, or he could never have invented a shorthand of his own in the first place—just let one of you try it: and he obviously didn't mean us to get his treasure without working like the devil.

"And then there are all sorts of other points to be considered. Even supposing that the message is in English, and not in Greek or Latin or French or Italian, you've still got to remember that a great number of words in use in 1695 are obsolete to-day, and a greater number have had their pronunciation altered. Since shorthand's entirely based on pronunciation, that doesn't tend to make things easier. Again, he may first have turned what he wanted to say into gibberish by using a simple cipher involving the substitution of one letter for another, and then put *that* into sound symbols. *Jasper Mauberley,* moving one along, becomes *Kbtqfs Nbvcfsmfz.* To make that repeatable you'd have to supply vowels: but you don't have any rules to guide you there. *Kabotikifess Nubavok-Fassimofes*—doesn't sound much like Jasper Mauberley, does it? And it's me that's got to make sense of the blasted thing, God help me, while you sit round on your haunches like a lot of stuffed fish and cluck with impatience."

"Fancy being able to do that in your head!" exclaimed Lina admiringly. "And anyway, fish don't cluck, Nunky. We'll all be as good as gold—bring you wet towels and strong whiskies, and cheer you up no end."

"Indeed you won't!" he contradicted, rather sharply considering that he was speaking to her. "You'll leave me in complete peace till I've got the hang of the thing, young woman."

"Very well," she agreed. "You're in Coventry till you've solved it—nobody speak to him, please."

"Then you think you *will* get the key all right, sir?" asked Hinkson, a little tactlessly, handing round the champagne.

"Of course, of course," averred his employer. "It's only a question of time and application—and experience, of course. Come along and make yourself useful."

Whereupon the party broke up. It was only eleven o'clock in the morning, and I think we all felt unexpectedly flat and uncertain what to do. After comparatively hard work for a full fortnight we were suddenly left idle; we could do nothing to help, and had been pointedly told not to hinder. Still, to keep our spirits up was the fact that the treasure seemed a great deal nearer than it had done an hour ago, and there was always the glittering ruby to remind us of that. Nevertheless, Adair didn't strike me as the man to minimize his powers, and if he said a problem was going to be difficult for him, then it must be an absolute snorter.

I said as much to Montague as we strolled off to the dining-room in search of a drink—neither of us cared about champagne before lunch. A little to my surprise he disagreed, and when I pressed him to give reasons he treated me more frankly than he had done so far.

"There's such a procedure as magnifying the dimensions of a task in order to receive the greater glory for achieving it," he told me, his face for once serious. I happened to be looking at him full, and noticed for the first time how deeply lined it was.

Under his rather protruding blue eyes the flesh was wrinkled and leathery, and pitted furrows stretched downward on either side of the long nose. Only the top of his forehead, where it curved over to meet his thin grey hair, was entirely smooth. I have said that he both looked and was delicate, his dark clothes hanging on him as if they had been made for someone half his size again, and the veins on the back of his tapering hands most clearly defined; but today he gave me the impression of being more than physically feeble. I felt somehow that his mind too was ill, or ill at ease. His words added to my surprise, and I didn't pretend differently.

"Is that like Adair?" I asked, and he shrugged. "You've known him two years, haven't you? I've known him forty, and they say that those who live longest see most—provided they keep their eyes open. Anyway, time will show. If he takes more than a fortnight over those miserable scraps of parchment, then I'll admit I was wrong, and ungenerous, and all the other harsh things you must be thinking. I'm getting old, I suppose—old and crabbed and mealy-minded. All the same, Adair has the instincts of the fisherman of fiction and so-called joke. His quarry's 'In capture huge, stupendous in escape'. And he will talk as if he knew everything and we knew nothing, too. Did you hear all that heavy stuff about shorthand being the representation of resentation of sounds by signs? I was tempted to ask him what he thinks ordinary print is.

"Still, let's change the subject. What's your opinion of that ruby?"

I laughed.

"The vast jump leaves me stranded," I said. "I thought it looked rather good, though I don't know a lot about precious stones."

"No? Well, I happen to, and you're quite right. It's a fine stone, worth four thousand pounds of anybody's money."

"As much as that?"

"Yes—unless it's a very clever fake, or has a flaw I missed. To be honest, I'm judging by the briefest examination: the inestimable Lina collared it, and I haven't seen the thing since. Have you ever watched a toad catch a fly?"

Again I laughed.

"But for Shakespeare I'd have called that ungenerous too," I said. "But I suppose lady toads have the same habits with their jewellery?"

"Well, ladybirds may not, of course. Am I to infer that you do or you don't?"

I knew at once what he meant, but queried it in order to avoid a *faux pas.*

"Like the inestimable Lina?"

"Just so."

"Well, I'll tell you if you'll tell me why you bother to ask."

"Agreed."

"Then I don't."

"Splendid—nor do I. As for my reason for raising the point, it occurred to me that I might have been indiscreet."

"In referring to her as a toad?"

"Yes—you might have passed it on."

"I shouldn't have dreamt of doing so: but would it have mattered particularly?"

"Indirectly: it might have caused a disagreement between Adair and myself. Please don't ask me to explain further."

The conversation then drifted on to less personal topics, but afterwards I found plenty to reflect upon in what had already been said.

IV

AT lunch that day Adair appeared for only a few minutes, gobbling his food with scarcely a word for anyone and then retiring to his study after a backward scowl at Hinkson, still eating. Later, I found myself engaged to play golf with Tilly and Lina and Buck. The little chap played, and amazingly well, making no secret of the fact that he had once been a caddy in America in his youth, before—as he put it—deciding that he could make better money in other ways. Lina, incidentally, was no snob—where credit is due I hasten to give it. Perhaps the significance of my remark about Judd's pocket and her handkerchief may already have made that clear, though. I'm sorry not to have brought the chauffeur properly on the scene yet, but you shall have him in due time.

The nearest course proved to be a moist nine-hole affair adapted to their own ends by rabbits—the furry kind: but we managed to get a game off it. In view of my handicap of five, which I don't play down to these days, I partnered Tilly, who was rather better than a beginner against the scratch Buck and the hopeless Lina. She managed still to be graceful, which may be accounted to her for a feat, since not many people can miss a teed ball five times consecutively without losing something in deportment: but it was her only accomplishment with a club apart from the boring of innumerable soggy grooves in the unprotesting turf. I grinned sympathetically at Buck from time to time, and agreed readily when presently he suggested that we should turn the outing into a four-ball match, he giving me a stroke at every third hole.

For the occasion Tilly seemed to have explored the depths of her wardrobe. She wore, believe it or not, a short moleskin coat above one of those flimsy Hungarian blouses embroidered with red and blue (and utterly wasted, be it said, since although thick-set her figure was incredibly curveless), a monstrous mackintosh skirt, conceivably home-made, openwork silk stockings, and green leather boots fitted with rusty zipper fasteners and reaching half way to her knees. It was a sight about which a mass-observer could have written a fair-sized pamphlet, I don't doubt. She appeared very keen to do her best, and did produce half a dozen passable shots, but otherwise her game was undistinguished.

As a rule she was a silent brooding girl with an air of intense concentration upon matters far distant in time and space, but today she relaxed a little, and even smiled once when Lina's ball vanished full pitch down a rabbit-hole. For the rest of the time her plump face maintained its usual gravity behind her steel-rimmed spectacles. Her hair, as always, hung almost down to her shoulders in a series of limp twists, and her nose shone pinkly. Her appearance afforded a great contrast to the smartness of Lina's, of which the main items were a blue suede jacket, sensible shoes, and a faultless complexion.

"Do you advise lifting the right heel clear of the ground at the top of the swing, Mr. Purdon?" she asked, after duffing her drive at the eighth.

"If it helps you to keep your balance," I answered, rather inanely. I've given up letting myself be inveigled into giving other people lessons about a game of whose theory I'm not at all sure.

"Ah, balance," she repeated. "Rather like instinct, don't you think? You either have it naturally, or you don't have it at all."

"Within limits," I agreed. "But everybody has to learn how to stand and walk—one isn't born able to do it."

"But may that not be due to lack of the requisite physical strength?" she objected. (It is a fair sample of her diction: inclined to be redundant, and curiously uncolloquial.)

"I'm afraid I don't remember," I said. "Do you mind standing back a bit? I'm going to swipe."

Afterwards, when I had done so successfully, I fancied from her expression that perhaps I had been a trifle short, and so tried to think of something to say that would show I hadn't been trying to snub the poor girl. She must have a thin enough time as it was, what with her father's frequent bullying and Lina's complete indifference, and there was no need for me to play the lout. Now normally I don't have much trouble in finding a subject for a casual remark, but on this occasion nothing would come to my tongue for several moments. I can't imagine why that should have been so. I wasn't shy of Tilly, or afraid of her, but there it was, and at last, in desperation at my own shortcomings, I asked a most uninspired question.

"I wonder how your father's getting on with the clue—do you think he'll be able to manage it?"

Her answer was not quite what I should have expected.

"I dare say they'll work it out between them—Mr. Hinkson is very good at that sort of thing."

An ordinary enough reply, you may think: but it wasn't, really. You see, as far as cipher work went Hinkson compared with Adair, in common repute, much as number seven of the average Varsity XI compares with Hammond as a forcing bat: likely to be far better than the first man who goes by in the street, but a great many thousand runs short of the original.

It occurred to me instantly that her father's secretary must have been cracking himself up to her. I knew Adair wouldn't: he never praised people behind their backs, least of all to Tilly, whom he hardly ever spoke to except to give an order or point out a fault; and I could think of no other reason for her implied high valuation of Hinkson's abilities.

"Oh, he's really good, is he?" I suggested,

"Oh yes," she assured me. "And, not only at ciphers, but at a variety of things. He said that the ruby that Daddy found in the box this morning was a Siamese stone, and probably worth over three thousand five hundred pounds."

"As much as that?" I murmured for the second time that day, though now with false surprise: the amount tallied well with what Montague had mentioned.

"Yes: and if it had been better cut it would be even more valuable. If that's just a token of the real treasure, then it's hardly possible to guess what that may not be worth—hundreds of thousands of pounds, perhaps. Then I might be able to—"

But here she suddenly stopped in confusion. We were walking slowly across to the others at the time: they had taken a different line of country to the hole, and lost Lina's ball in doing it. Unconsciously Tilly had slowed her steps with her words, and was now looking rather unhappy, as if she had been caught repeating a secret to herself.

"Yes?" I prompted. "Be able to do what?"

Perhaps it was the unaccustomed fact that somebody should be showing a definite interest in her, or in anything she might be about to say: I don't know, only that the next few minutes were a revelation to me.

"Have a proper dress allowance!" she burst out abruptly, her weak eyes intent on the sky-line and her cheeks flushed. "Not to have get myself up like a scarecrow in other people's throw-outs."

She gave her mackintosh skirt a savage slap.

"Mrs. Darling, Edgware Road," she said bitterly. "Likewise the blouse and these bloody boots. Other people's filthy clothes all the time because I can't afford anything new, anything decent, anything clean, even—because I have to make do with what she sends me, all the old rubbish that nobody else would touch without gloves on. Two and ninepence this skirt—three shillings the blouse. The boots were dear—they cost half a crown each. That dress I wore yesterday was the same, and dear at that with all the stuff gone under the arm. And what you can't see is worse. I'd sell anybody my underclothes for five shillings and be in pocket, stockings and all."

She paused, and with a jerky movement took off her spectacles.

"Now I can't see you," she said. "I can't see anything without them. You're wondering why I don't go to Marks & Spencer's rather than Mrs. Darling, aren't you? Why anyone who wasn't stone blind would spend a farthing on such foul things as these boots. Well, I don't get the choice, that's why. A cheque for thirty shillings every month made out to her—that's what I dress on. She gets the cheque and I get what she sends in return—all the vile things you've seen me in, and some I wouldn't ever dare wear. Can you wonder I don't bother about what I look like? That I just stick on something different from what I wore yesterday?

"Oh, I know I'd never be worth looking at twice, and maybe I don't make the most of what I have got, but I simply haven't the heart. I don't care—I just don't care. She's a horrible woman with dyed red hair and a sly leer, and I don't know which I loathe more, her or her second-hand clothes. And last time she had the—the absolute impertinence to put in a pair of some schoolgirl's old gym knickers, all darned and shiny. Christ, sometimes I don't know how to bear it all, Tony. But I did draw the line at that. I got hold of the two legs and pulled them in half and sent them back and told her to throttle herself with them. And then, because I only had eightpence of my own, I—oh well, it doesn't matter telling you now—I stole half a crown from

Lina's bag and bought myself something clean. You see, I'm a thief as well as a scarecrow.

"Oh dear, I'm sorry—they're waiting for you to play. I didn't mean to say all that—please forgive me."

But it was I who should have sought her forgiveness. Just now I used the word lout, and I give you my word that at that moment I felt more utterly loutish than ever before in my life, or ever again, I hope. I had actually grinned to myself about her clothes, and not had the wit to consider that there might be some good reason for her wearing them. I said what I could to comfort her before we rejoined Buck and Lina, and for the rest of the afternoon was quite off my game.

Adair came in to dinner that night, and I heartily wished he hadn't. I found myself staring at him speculatively when he wasn't looking, and trying to square my new knowledge of the man with previous conceptions and experience. That he was mean in many ways I had already known: but not that he would descend to such depths of meanness as to dress his daughter on thirty shillings' worth of cast-off clothing a month, when he must have thousands in the bank. I didn't for a moment doubt Tilly's story: if ever a girl spoke from the heart, with utter truth, then she had done so that afternoon.

That she chose a comparative stranger to confide in wasn't quite so remarkable as it appeared on the surface: there must have been very few people whom Tilly could think of as other than strangers. She had momentarily come to the limit of endurance; she had had to say something or scream, and she had preferred to speak. It was always possible, too, that she had been guided in her choice of audience by the fact that I had never been openly hostile towards her, not shouting nor snubbing nor noticeably avoiding.

So I looked at Adair repeatedly, and wondered. It was understandable, if only just, that although Tilly was his own daughter he might greatly prefer Lina.

It was even understandable that because of this preference the latter would always get the tit-bits and Tilly the scraps. What I simply could not fathom was the attitude of mind which enabled him with apparent naturalness to treat one as a princess and the other as a not too clean servant girl. I regarded his white thatch of hair, his striking face with its aquiline nose, military moustache still flecked with brown, and firm chin, his general air of gentlemanliness, if I may use so out-moded a word, and sought for the clue to his behaviour, the tell-tale sign which I must for so long have missed.

I remembered what Roger Montague had hinted that morning: at a vanity beyond the ordinary, of the kind which would invent difficulties in order to be able to boast of having surmounted them. Would that explain his treatment of Tilly? I hardly thought so—it didn't seem to fit at all.

He was talking to Montague about his own clue, and I followed his words vaguely.

"... a preposterous idea. Why you should always imagine the worst beats me, but you always do. What have the others got to say? You, Tony?"—suddenly turning in my direction.

"I'm sorry," I answered. "I didn't entirely get Mr. Montague's meaning."

"Pooh—simple enough! He suggests that the parchment is a hoax, nothing to do with Jasper Mauberley at all, but the work of Boon, the man whose diary I translated. He bases his ridiculous suggestion on the coincidence that both documents were written in private shorthands. I say that the presence of the ruby proves him wrong—what's your opinion?"

"Well, I don't see much point in planning a practical joke that can't take effect till you're mouldering," I said. "Practical jokes are perpetrated chiefly to give the author a laugh, not a lot of strangers heaven knows how many years in the future. And anyway, I should have thought the argument could be settled easily enough."

"Settled? How?"

"By expert advice,"—and as I spoke I grinned at Montague, recalling what we had said about experts on the first day of the search. Apparently he too remembered, for he grinned back.

"Mauberley was writing seventy years or so before Boon, wasn't he?" I went on. "Then why not cut off a snippet from the rigmarole you found this morning and have it vetted by somebody who knows something about papers and parchments. You'll probably find it could be dated pretty accurately."

He received the idea graciously.

"Not bad, not half bad," he conceded. "Only a negative proof, of course, but probably simpler than an ink test—I'd already considered having that done."

(This, I felt sure, was a lie. He hadn't given the matter a moment's thought, but wanted to make me believe that I hadn't by any means bested him.)

"I still maintain that the ruby answers the question out of hand, though," he added. "And where is the ruby, by the way?"

"I've got it," said Lina at once, with her most kittenish smile. "I'm looking after it for you—it's far too pretty to be stuffed into some old pocket and forgotten. Or do you want it back?"

"Oh no—keep it for me. I'm too busy to play with pretty stones. Finished, Hinkson? It's time we were getting back."

"Oh dear, that means no bridge," murmured Tilly, half to me and half to herself. She was keen on the game, and good at it, and most evenings Hinkson and Montague and I helped her to make up a four; except on the rare occasions when Adair wanted to play, when one of us pleaded letters to write. Lina was no more use at a card table than on a golf course, and Buck was never indoors after six o'clock or so. He spent the hours of darkness prowling round silently outside, lying in wait for an intrusion which I don't think anybody seriously expected.

"Bridge!" echoed her father scornfully—he had sharp hearing. "Here we are half way to fame and fortune, and all the girl thinks about is a twopenny-halfpenny game of cards. Pooh! Much more to the point if you did something useful. I've an old coat upstairs with the lining out—I'll send it down."

For a moment I thought Tilly would rebel. Her face flamed violently, and she seemed about to speak; then changed her mind, and turned away with a despairing shrug.

V

LATER I decided to go out for a walk by myself. I didn't feel like reading, I preferred not to write to Beale until I had digested the day's events, and there was nobody to talk to. Montague had gone to bed, complaining of indigestion, Tilly had retired to her room with an armful of Adair's mending,

Hinkson was busy in the study, and Lina was not my choice for present company at any price. She did make an attempt to dissuade me from going out, but I laughed it off.

"It'll be terribly cold," she said. "You'll freeze—much better snuggle up by the fire."

She was in the communal sitting-room following her own advice, a magazine on her elegant silk-clad knees, and as she spoke she drooped her long dark lashes at me, cocking her head a little to one side and giving me to understand—I believed—that I need not do my snuggling alone. When you consider that she knew very well I disliked her after witnessing the incident with the cat and the paper bag, you will perhaps agree that she was a resilient young woman, the thickness of whose skin was not to be judged by its rose-bud bloom. She was wearing a claret-coloured silk gown cut low over her seductive bosom and high at the back, so that except for the excellent way it set on her one might easily have supposed that she had deliberately donned it the wrong way round. (I should have said earlier that we dressed in the evening: a fad of Adair's with which one fell in without quite seeing the necessity. I've no particular objection to a dinner-jacket and a boiled shirt, but a very marked one to the point of view which insinuates that I am temporarily a bounder after seven o'clock in the evening if otherwise arrayed.)

Sitting there by the old-fashioned open fire-place, in which blazed a huge log, Lina certainly looked attractive, and if I had been ignorant of her nature I might possibly have been taken in, and regarded her invitation as one of fate's better contrivances. As things were, I would as soon have nuzzled cheek by jowl with a well-groomed female boa constrictor.

"Thanks," I said: "but you'd really better find some one nearer your own age."

"Such as?" she queried idly, not moving nor showing signs of minding much one way or the other.

"Oh, a nice young curate or a dozen Boy Scouts." She hardly smiled.

"You must be muddling me up with Tilly," she said. "What would I do with a lot of Boy Scouts?"

"I can't imagine. Enter a pack of Wolf Cubs; pause; exit a pack of were-wolves; *manet Circe.*"

"Who ate who?" she asked, but I didn't stop to explain.

She had been right in her forecast of the weather conditions. The night was dark and cloudy, with neither moon nor stars visible, and the slight wind was bitter. The flag-stones beyond the main entrance—an imposing porch that terminated in a solid weather-beaten arch—glistened with frost when I shone my torch on them, and I had to mind my steps. Well wrapped up, and with a flask in my pocket, I picked my way past the desolate out-houses, skirted the barren orchard, and passed out of the gardens proper along a narrow path that led to a small ash wood. Everything was very silent, the only sounds being the contact of my feet with dead frosted leaves and the occasional rustle of an unseen animal or bird. The air smelt unaccustomedly sharp once I had left behind me the obnoxious scent of a smouldering bonfire.

Perhaps you will expect to be told now that I directed my thoughts to ultramundane things: greedily drank in the crisp night air, communed with invisible nature, and performed similar conventionalities. In sober fact, however, I walked fairly steadily for about seven minutes, shivering in spite of overcoat and muffler and thinking of Mrs. Darling and her contaminated wares, and then stopped to tell myself that I was a fool not to be in bed, or at least indoors. I then fumbled in my pocket for a cigarette.

I realized, now I was out there alone, that what I had really been wanting was not congenial company but a period of solitude, during which I could perhaps brace myself for a task that I must soon face. In view of Tilly's revelations that afternoon, and their consequent effect upon my estimation of her father, could I continue to stay at Mauberley Grange without loss of self-respect? And if I did stay, was there the slightest chance that I could make things easier for the wretched girl? If not, wouldn't it be better to pack up and go rather than make myself feel small and hypocritical by pretending that I felt no differently towards him?

These questions were awaiting my attention: to answer them was my imminent and inescapable task, and as soon as I felt refreshed enough I would try to do so. Meanwhile, I repeat, I sought an interim of mental blankness, and this is far more easily achieved for me in darkness than in light, and in discomfort than in luxury.

Unfortunately, Buck chose the moment when I was about to light my cigarette to stick something hard in my back and bid me remain still. He laughed quietly when he saw who I was, and seemed a little disappointed. I congratulated him on his stealthy approach, agreed at his request not to smoke, and shared my flask with him. I also abandoned all hope of solitude.

"How do you manage to creep about like that in the dark?" I asked. "Natural ability, or years of practice?"

"I dunno," he said. "A bit of both, I reckon. It's easy enough to put your feet down light if you lift 'em up first and take a bit of care. And I wasn't as quiet as all that, anyway. You must have been dreaming, weren't you?"

"Perhaps—I got fed up with being indoors."

He said nothing to that for a moment, and then asked a question.

"D'you know what I get paid for doing this job, Mr. Purdon?"

"No, I'm afraid not—ought I to? To tell you the truth, I never thought about it."

"No, I suppose you wouldn't. But I have, you bet. I get eight pounds a week and the stuff they call food. Now what does that make you say?"

I looked at him, but could distinguish nothing of his features. He was merely a blurred presence beside me: yet I fancied his bright eye perked enquiringly.

"Well, what sort of an answer do you want?" I parried. "Polite, honest, or none at all?"

"Oh, let's have a bit of truth for once. Not meaning anything special against you, of course: but what do you think?"

"All right: it strikes me that you either get too much or too little."

"Yes?" he encouraged.

"I mean, too much if there's nothing definite and important to watch out for, and too little if there is."

"Yes?" he murmured again—we were speaking in low voices. "Make the last part clear—I get the first okay."

"Well, if this fellow Warner is really in the background, waiting for the Major to locate the treasure so that he can burst in with a gang of toughs and collect same, I don't quite see what you'll be able to do about it. Of course, I don't know your capabilities in that line, so maybe you could do it, and nothing short of a machine-gun battalion would get the better of you. All I meant was that if you're that sort of chap, then you're worth a sight more than eight pounds a week."

"Ah, now I see—thanks for explaining. Well, I reckon I can tackle any three other men you care to name, and five if you take their gats away, but not more unless they're cripples. Now I'll tell you what I've been thinking: that so far it's been money for jam, and I'm wondering if things'll be different now."

"Different? Why should they be?"

"Now be yourself, Mr. Purdon, be yourself. If the old boy never gets a step nearer his treasure than he is today, he's still a pretty fine ruby up, ain't he? One that'd be worth reaching your hand out for on a dark night."

"Oh yes, of course—I was forgetting. But wait a bit: how is anybody outside the house to know anything about it? As far as I can see, they won't."

Buck laughed softly.

"Now you're getting warmer," he commended. "How is anybody outside the house to know? All right, I'll tell you how: the same way as Warner's to know when the treasure's found. How were you reckoning that would happen? Will Adair stick an advert on the front page of the *Mail?*"

I said nothing for a little, considering the import of his words. Then I offered him another drink, and took one myself.

"It hadn't occurred to me before," I explained, "because I can't pretend I've ever taken this Warner chap very seriously, but there's only one way that I can see. If anything's going to happen, then he's got a spy in the house."

"That's it—you've hit the bull, Mr. Purdon."

"If there is a bull," I reminded him.

"All right, if there is one. But what did you mean by not taking Warner seriously? Don't you believe the diary was photographed, and there was somebody wanted to buy this place?"

"Oh yes—about the diary, that is. The other I've always put down as a trick of the agents."

"Well, maybe so, maybe not. But let's suppose not. Let's suppose that Warner's outside there somewhere waiting. Now what'll you say if I ask you who's the spy?"

"I'll say I don't know: but I could bear to start thinking about it."

Again he laughed.

"You told me you were fond of amateur detecting once, didn't you? Well, have a go at it. Where do you start?"

"By leaving myself out—provided you agree, of course."

"I guess so—if not, you'd be about the last person I'd be talking to. And you can leave me out as well."

"Right: that makes about seven to consider, not counting the Major, because he'd hardly split on himself, and Mrs. Judd, because she doesn't strike me as a very likely conspirator."

"Not her—she don't dare sneeze without permission, I reckon. All right, seven: the two girls, Hinkson, Andrews, Judd, the cook, and Mr. Montague. Personally I think you can leave out the cook too: she ain't got the brains to be nobody's spy."

"Very well—you probably know best about that. Likewise the village wenches? We forgot them."

"Yes, they're simple too. Then there were six. Who's your choice?"

"Give me time, man—I'm not clairvoyant. Let's see if we can't reduce the list a bit, first. What do you say to leaving out Tilly?"

Rather to my surprise, he was inclined to demur.

"Her pa don't like her—she don't like her pa. Quits, as you might say. I guess I'd rather leave her in."

"As you wish, though I'm afraid she won't be my pick. I don't think she's quite the sort to approach the enemy and offer to help them, and I don't see the enemy choosing her as a likely person to square."

"Maybe not: but don't go forgetting she knew Warner. I'm not saying there was anything between them, mark you—only that they had an opportunity for getting pally. Still, she wouldn't be my first choice, either."

"Then suppose you tell me who would be?" I suggested.

"All right, I don't mind. In order, Hinkson, Lina, Montague, Andrews, Tilly, Judd."

"You've evidently been worrying this thing out," I said. "Why put Judd so low down, by the way?"

" Because I don't reckon he knew enough to be much help to anybody. He don't even live in the house, and a spy wouldn't be much use if he couldn't hand over a plan of the rooms and passages. Warner and crowd don't want to be told the treasure's in the fourth room on the second floor facing east. They want a clear idea of how to get there quickest, and who they'll pass on the way."

I felt justified in disturbing his complacence.

"Provided there is a treasure," I told him, "I'll lay you five to one it's out of doors and not indoors."

"Why so?" he asked sharply. "Hinkson said the old codger who owned it was bed-ridden."

"But not all his life. The idea is that he hid his treasure before he went to bed. Hence the moral certainty, according to the Major, that the clue would be in the house and the treasure not. Anyway, the first part worked out all right."

"Ay, that's true enough. I guess I've been a bit previous, thanks to little pretty-boy."

He remained silent for some three minutes, obviously pondering my information.

"Judd spends more time in the gardens than anybody bar me, don't he? I wonder. But then, that mightn't be much help—things change a sight more out of doors in a hundred years than they do in. I know these secret clues—read about dozens of 'em in books. 'Take ten paces due north from the shadow of the old apple-tree by the sundial, then seven paces south by west, then dig till your braces bust.' And when you come to look, you find your old apple-tree's been firewood before your great-granny knew which side of a bed was safest, and you could bust a million pair of braces and find sweet Fanny Adams. If this durned treasure's out of doors, then it's going to be perishing hard to find—that's my guess."

"I don't doubt it," I agreed, and kept my thoughts to myself. Jasper Mauberley would scarcely have gone to so much trouble in hiding the clue, I reasoned, nor in reducing it first to shorthand, if he had not intended his treasure to be found at all; but he must surely have been clear-sighted enough to realize that this might not happen for generations. I was hopeful that if Adair managed to unravel the clue, it would guide us by more enduring marks than sundials and the shadow of a tree.

I turned towards Buck, and was about to ask why he gave Hinkson pride of place on his list; but he spoke first, once more surprising me. He was an unaccountable little man at times.

"Well," he said, "I reckon I must be off on the prowl. Think it all out, Mr. Purdon, and let's hear about it some time. And it wouldn't be a bad idea to keep this between ourselves, would it?"

"A very good idea," I concurred. "Don't fall asleep and get frozen."

I hurried indoors as speedily as I could, for I was shockingly cold. In my bedroom I crouched as near the fire as was possible without scorching my clothes, and for some minutes considered the question of an imaginary spy's identity half-heartedly, until I fell asleep. I roused hours later cramped and wretched, and tumbled into bed un-washed, which means that I couldn't have been more than half awake. Yet, all the same, I had one clear thought running through my head. 'There's going to be trouble soon ,' I told myself. 'Trouble, my boy—just you see.'

And thus, though nobody will believe me, I also have lain down among the prophets.

VI

DURING the next two days Tilly noticeably avoided me. I think she was worried about her outburst during the game of golf, and afraid that I might tell her father. I had not thought to assure her that I would keep her confidences to myself, nor she apparently to ask as much. By my manner I tried to show her that she had nothing to fear, but judged it better not to refer to the matter openly. In this I may have been wrong: the fact remained that she avoided me as far as possible. To be frank, those two days were pretty dull. Adair was scarcely to be seen at all, and when he did put in an appearance at meals he never stayed more than ten minutes or so. It was evident from his manner, and such few remarks as he let fall, that he was not getting on particularly well with the deciphering of Jasper Mauberley's clue, and Hinkson told me without hesitation when I en-quired that this was indeed the case.

It was after dinner on Monday night, and—as he put it—he was snatching a little fresh air before returning to the grindstone.

"Though for all the good I'm doing I might as well go to bed," he added with his bonhomous smile.

"Oh, nonsense," I said. "Tilly tells me you're by way of being an expert yourself at ciphers."

"Really? That's very nice of her; but, you see, the thing we're on isn't a cipher, strictly speaking, and you could write what I know about seventeenth-century private shorthands on the proverbial threepenny bit, and still have room for a prayer or two."

"Then what do you do all day long? Sit and hold the Major's hand?"

"Hardly. I'm given donkey work: making endless copies of the beastly message, classifying the different signs in order of frequency, and peering through a magnifying glass to see if they were made with an upstroke or a downstroke. I shall have to start wearing specs if it goes on much longer."

"And will it?"

"Heaven knows—he hasn't made any sort of a beginning yet."

"But once he does, I suppose it'll all be plain sailing?"

Again he smiled, this time a trifle condescendingly. Whatever he might say, compared to me he was an expert, and he knew it.

"That depends on the point of view," he told me. "Relatively plain, shall we call it? About three days' really hard work, so the Major says. You see, even all that time ago they knew the tricks of the shorthand writer's trade—leaving out the vowel marks and so on to save time, or through laziness or sheer devilment. In Jasper's case it would probably be the last. It's not always too simple when you're confronted with an outline that might equally well be 'had' or 'head' or 'hid' or 'heed' or 'hide' or 'hoard'."

"So you have to go by the context," I suggested, to show I didn't want to be treated as an infant. "As a matter of fact, I write a kind of Pitman's myself. Doubtless the inventor wouldn't recognize it, but at times it serves its turn. And when the half dozen previous words are all in the same state of uncertainty, you have to sit down and consider the various possibilities. 'Bury or burrow or borrow the gems or jams in the wall or well at nine or noon.'"

"I say, that's pretty good on the spur of the moment!" he cried, a shade too heartily. "I'll have to get you roped in on the job."

"Oh, please don't—I should hate it. By the way, to change the subject, do you happen to be able to describe your predecessor's appearance? I'm rather thinking of running into Gloucester tomorrow if I'm not wanted to tie up sacks of diamonds, and I thought it might be as well to keep my eyes open. I did ask Adair once, but he was a bit vague—gave me the impression he'd never really noticed the chap."

"Why, yes, I can give you a sort of idea. I saw him for a few moments once when he came to collect something he'd left behind. He was a shortish sandy-haired man of about forty-five, with a small moustache and an Old Harrovian tie. I can't say I cared much for the look of him. But surely it's most unlikely that you'll see anything, isn't it? If he is round these parts, he'll be lying pretty low."

"Undoubtedly: but Gloucester's fifteen miles off. Anyway, it was only an idea, and probably a wild one."

(In fact, of course, I was trying to find out if Hinkson would admit to knowing Warner's appearance. That he did so willingly made me think that Buck must be wrong in his suspicions. It wasn't till I had been alone for half an hour that I realized I had committed myself to visiting Gloucester the next day, wet or fine. I didn't feel inclined to let Hinkson suspect I had been pumping him.)

Later that evening I wrote to Beale again, having by now definitely decided that I would stay on. I don't quite know what made me do this, but I think it was my talk with Buck in the woods, combined with my subsequent momentary foreboding of trouble to come.

My first letter, the one with which I began my share of this book, had elicited a short note saying that the party sounded interesting, and asking for any further news I could supply.

I think he said that more out of politeness than anything, but the opportunity was too good to be missed. If my torpid presentiment should by any chance come true, then there might be work for the police at Mauberley Grange. It may perhaps be held that I had no good reason for supposing that anything untoward would happen: in which case I have failed to make it clear that the other members of the party were a very mixed bunch, and that I didn't particularly care for any of them. And when all's said and done, with a treasure in the offing you can't be too careful or wildly speculative, even if the treasure does belong to somebody else.

But, it may be objected, I have scarcely done more than mention two of the party: Judd and Andrews. The latter wrote left-handed, or pretended to, and the former dropped Lina's handkerchief in circumstances—the envelope—which made me harbour base thoughts. So what?

Well, to deal first with Andrews: he was nothing to look at, thin and grey-haired and about Hinkson's height. The most striking thing about him was his quietness of foot, upon which I am sure I have already remarked: but one naturally expects a butler not to shuffle. I don't suppose I should ever have thought twice about the man, though, if I hadn't happened to see him signing for a registered letter one afternoon. The postman held out a pencil, Andrews took it in his right hand, put the slip of paper against the wall at a convenient height, was about to sign his name—and then suddenly took the paper down again and changed the pencil into his left hand. The postman was looking elsewhere at the time, sneering at the state of the garden, probably, and only I saw Andrews' action; and it made me think. It was impossible to doubt that his natural instinct had been to use his right hand, and that he had deliberately if tardily chosen to employ his other, and while there might be perfectly simple and straightforward explanations for such a change, there might also be some of the other kind. As well, it seemed to me that the way he glanced sharply sideways at the postman to see if he was observed held something of furtiveness. I say no more, except to repeat that it made me think. On the whole, though, my suspicions of Judd were of a deeper shade. Now, it is a commonplace that there are two kinds of male ugliness, the attractive and the repellent. I speak of attraction and repulsion in terms of female opinion, of course: but the matter isn't quite so simple as it sounds. I have heard Victor McLaglan and Wallace Beery enthused about by one girl, who found them equally fascinating; condemned by another, who said she would rather look at a couple of real apes; and differentiated between—differently—by a third and fourth. Again, it is very doubtful if any given man will bear out any given woman's view of a second man's looks. Sometimes the reason for this discrepancy of verdict is vanity, or jealousy, and sometimes genuine diversity of taste; but the fact remains that only in relatively few cases do a majority of both men and women agree in such matters. In a book by Havelock Ellis called *The Criminal* there is a series of sketches depicting the heads of various convicts as drawn by Dr. Vans Clarke, a former governor of Woking Prison. They may or may not have been good likenesses, but I never met anyone who could honestly say that he found even one of them remotely attractive. That is the sole collection of faces I know of, however, which obtains from those who look at it a uniform expression of opinion.

I'm being long-winded. What I'm trying so laboriously to say is that although I myself found Judd's features almost revolting in their gross coarseness, I didn't discount the possibility that other people might be found to disagree with me: in particular, Lina, from whom I should never have expected a high standard of taste. The quality of maleness would be more likely to interest her than the quality of handsomeness, and Judd was aggressively male: tall, broad-shouldered, long-legged, thick-wristed, hirsute.

Perhaps it may be thought that I had a down on the girl, and was determined to see no good in her, even if I had to poke my eye out in order not to. I can't defend myself with much hope of success, so all I will say is that I tried to be fair to her. Judd may conceivably have found her handkerchief in one place, and one of her discarded envelopes in another, have combined the two discoveries with the intention of returning both, and then have lost the combination by accident. If so, however, I would have wanted at least a bishop's evidence before I believed.

Put it down to the unpleasant nature of my mind if necessary, but I shouldn't have betted against some sort of intrigue between the two. I think I've said that Lina did very little searching after the first few days, and that Judd scarcely ever came into the house. He was supposed to be tidying up the garden a bit.

But how does that square with my hint in the first letter to Beale that I also thought there might be something between Lina and Hinkson? Well, there I admit I wrote without much to go on. All I had was that several times I fancied I had caught them exchanging glances which seemed concerned not so much with the present as with the past: the kind of look which doesn't say 'How does that strike you, I wonder?', but 'I wonder if you remember that I said much the same thing last Tuesday?'

One thing I was certain of, though. If Judd had ever loomed larger in Lina's life than one's adopted father's chauffeur should, then Hinkson neither knew nor suspected this. He had far too good an opinion of himself to put up with competition from anyone so markedly his inferior in looks and standing.

As it happened, the very next day produced indirect support of my theories about Judd. I duly set out for Gloucester about half past ten; alone, as I thought, because Montague wasn't feeling well, and Tilly felt she oughtn't to leave her father. (The truth was more probably that she preferred not to be alone with me in case I embarrassed her, though I shouldn't have done, of course, but I will concede that she may not have felt sure about it.) Lina, I need hardly say, I didn't dream of asking to accompany me, and anyway she wasn't visible.

Fate, however, always carries a knuckle-duster. I had reached the bottom of the cart-track called a drive, and was slowing down to take the sharp left-hand turn into the road. On my right, twenty feet away, stood the lodge, a small stone-built house considerably more modern than the Grange itself, but in worse condition. Owing to the angle at which it faced me, I heard nothing until I was almost past: and then came the unmistakable sound of female voices upraised, clear through the open driving-side window.

This startled me a little, because—as I have said—Mrs. Judd, who inhabited the lodge with her husband, was a most feeble ironed-out sort of creature whom I should have thought quite incapable of shouting: yet the louder voice reminded me somehow of hers. I looked, and discovered that it was, and what I then saw surprised me even more. Out of the porch came Lina, backwards, and urging her on followed the chauffeur's wife brandishing an umbrella. Her bedraggled hair was screwed into a kind of bun on top of her head, her usually white face was full of colour for once, and she seemed to have been weeping.

"Get out!" she yelled, in a strange hoarse rasp. "Get out, you stuck-up little slut! And don't you come monkeying round 'ere no more—you and yer so-and-so half-crowns! Oh!"

The final exclamation was caused by sudden awareness of my car and staring visage. Lina followed her gaze, turned back, said something quickly which I couldn't catch, and ran towards me.

"Let me in!" she panted. "Did you hear? She simply screamed at me! Where are you going? I can come too, can't I? What a nerve! What a stone-cold perishing nerve!"

I was thinking on rather similar lines myself, though not about the same person. However, I could see no way of dislodging my unwelcome passenger short of employing violence, and so accepted fate's deplorable foul without protest.

Lina appeared pleased with the idea of visiting Gloucester, and was already undoing her crocodile handbag in search of lipstick and the rest of her portable beauty-parlour.

"And what roused the meek Mrs. J. to wrath?" I asked presently. "I couldn't hear as well as I should have liked, but it sounded as if you'd been trying to pass her a dud half-crown."

The girl laughed coolly: she was herself again. "I was going for a walk," she said, "and remembered I hadn't any money with me, so I went in and tried to borrow some, and she just blew up. God knows why—p'raps it's a sore point."

"I see," I murmured: though I didn't at all. The explanation sounded to me quite improbable: Lina wasn't the girl for a winter-morning walk, especially in flimsy shoes and no hat. If she had reason to prevent Mrs. Judd from making a scene about something unpleasant, though, she might just possibly snatch up her coat and bag and run down to the lodge, and I remembered having noticed Judd himself hanging about the house half an hour before.

In Gloucester, for Hinkson's benefit in case he should ever get to hear about my movements, I made some pretence of keeping a look-out for Warner. I said as much to Lina, questioning her about his appearance, but she seemed to think the idea fantastic. "He was a wretched little tick," she said. "He wouldn't have had guts enough to rob a baby's money-box."

"Why—did he try opening yours?"

"No fear—I keep my money where it doesn't rattle. And that reminds me—you'll have to lend me some. Now I'm near some decent shops I may as well buy a thing or two, and as I told you I've come out without a sou."

"How much?" I asked, as reluctantly as possible: I thought she might at least have said please.

"Oh, I don't mind—a fiver? Thanks—I'll pay you back, of course."

"Of course," I echoed, pessimistically. Only one incident stands out in my mind during the time we spent in Gloucester; or two, if you count seeing a daring old man on a tricycle try to cheat the traffic lights, and get spluttered at by a nervous policeman young enough to have been his grandson. It happened early, and it happened entirely by chance. We had agreed to separate for an hour, and then meet for lunch at 12.30, and for want of anything better to do I wandered into a bazaar. It was likely to be warm, and I was cold, and temporarily off beer.

In order that I shan't be accused of suggestiveness, I relate the facts with the minimum of elaboration. A display of magazines caught my eye, and I was wondering whether *The Modern Detective Monthly* (apparently featuring the exploits of seven hooded men and one half-naked girl) could possibly be worth reading for sixpence when I heard Lina's voice nearby, behind me and a little to my left.

"I'll have these," she said. Naturally I turned, looked once, and departed forthwith without her spotting me. One and elevenpence of my fiver had just been expended upon a pair of black silk bloomers.

In view of the weather I could understand everything except the colour, and the only rational conclusion I could come to about that—not that I bothered much, to be fair to myself—was that her purchase would eventually find its or their way into Tilly's possession unless they were to be an unseen sign of mourning for Mrs. Judd's complaisance.

There is little to report about the remainder of the day. We went to a cinema in the afternoon to see Paul Muni and Louise Rainer in 'The Good Earth'. I thoroughly enjoyed the film, but Lina voted it dull, which one might have expected from her. After tea we returned to Mauberley Grange, and it struck me in the car that she was unwontedly subdued. She might almost have heard suddenly about the death of a valued friend, if she had been the sort of person at all likely to have one. At the time I put it down to the effect of the picture.

I deliberately avoided Buck that night: somehow I hadn't managed to give the question of a possibly spy in our midst the attention I felt I should have done. Instead I went up to see Roger Montague, lying in bed reading Duhamel's *Salavin*—my copy, which I had lent him.

"Yes, I like it," he said, in answer to my query. "Yet it's disappointing, in a way. He tells you what the man did, and as often as not why he did it, but he doesn't make me *understand* why. I never for a moment identify myself with Salavin as I do with Albert Grope or Adam Jeffson or Michael Fane. I pity him, but I don't sympathize—if you won't think I'm playing with words by saying that."

"Oh no—I get your meaning. You begin the book in the belief that for once you're really going to examine a man with both eyes, and you finish by realizing that one's been closed for you forget how long. Anyway, I did. How're the innards?"

"Still a bit rebellious, thanks. I envy you leather-lined fellows with your feet on solid ground—it's no joke to live in a nearly perpetual swing-boat condition, abdominally speaking. However, don't let's be medical. What have you been doing to speed the undertakers?" I gave him the main outlines of my day, and said I hoped Adair would soon solve the clue, as I didn't relish the idea of hanging about for long in total idleness.

"Oh, idleness is all right," he assured me: "only I agree that this isn't quite the atmosphere for the carefree life."

At that I regarded him sharply.

"Which means exactly what?" I asked. "I'd rather like to know."

He smiled momentarily, and took a sip of barley-water.

"Isn't it simple enough?" he countered then. "Why are we all here?"

"Well, to look for the treasure, I suppose."

"And why are we looking for the treasure?"

"Well, because Adair invited us to."

"And why did we accept his invitation?"

I began to see what he was driving at, but couldn't altogether accept his point of view.

"You mean we've all got a little secret label on us—'Covetousness, his mark'? But I don't think it holds good for everyone."

"You being the first exception, naturally," he murmured with a smile, though not offensively. "Perhaps you're right, and I'm sure I hope so, but I'd like to hear you defend some of the others."

I declined the task.

"Accusation before defence, always," I said.

His answer was unexpected.

"No man can be compelled to incriminate himself," he reminded me. He need have said no more, of course: I would have changed the subject at once, or been willing for him to do so; but—perhaps characteristically—he went on to do what he had just said could not be required of him.

"Oh yes, I'm broke to the limit, or very near it. I was pretty well off once, but my bank manager would tell you a different story today. For one thing, I had two years' illness, and you've no idea how that runs away with the money. Then on top of that certain investments went wrong, and sundry other calamities beset me, so I can't escape the little label myself. I live in hopes of at least having a share in the finding of the treasure, and so possibly a share in its value. My eyes positively glisten with pecuniary desire every time I think of how much even one per cent might come to—can't you see them doing it?"

"Not unduly," I said, uncertain if he were ragging.

"Ah, but that's probably because you'd rather not," he told me seriously. "I make no bones about it: I could very well do with a few thousands. That's why I can't afford to offend Adair, and also, incidentally, why I pretend I don't believe in the treasure much. Possibly, too, that's the reason I see the labels so clearly on other people. Not meaning you, though: frankly, I haven't been able to justify your presence here at all. But take Hinkson: half his mail consists of bills, if I may speak in confidence."

"Whose?" I asked bluntly. "His?"

"Oh no—credit me with the fragments of decency, please. The nature of his correspondence is evident if you take the trouble to examine first its form and then his expression as he stuffs it unopened into his pocket. Again, perhaps I'm attuned to his feelings by experience: but I know what it's like to be dunned, believe me. Or take Lina. Regard her with

an unprejudiced eye, if that's possible, and ask yourself
how long she'd stop where she is if she had the means to
get away. I never look at her without thinking that her
spiritual home is a smart bar in Paris or Cannes or Vienna,
surrounded three deep by elegant foreign counts carrying
boxes of whatever those chocolates are they do the slinky
drawings of in the papers. Or else astretch on a sunny
beach, brazen in a brassiere. I can't bring myself to believe
she winters in England from choice.

"Tilly, perhaps, may be excepted, though I fancy her
pocket money would fit none but a very small pocket. But
what of Buck, with his bright eyes and capable hands and
his dull job? You've only to look at him to see that he's a
man of action, not a dreamer; that he'd rather be in motion
than at rest, and preferably violent motion, concerned with
firearms and what-not. When a wolf pretends to be a
watch-dog you can usually depend that he's doing it from a
wolflike motive. He may lick the hand that feeds him with a
dry bone, but only till the butcher's van arrives.

"In short, Tony, we're a rum lot, and most of us are on
the make, in my opinion, Adair included. Apart from his
determination not to be beaten by a dead hunchbacked
miser, there's also a distinct interest in the material side of
the affair, unless I'm much mistaken. When he told me the
price he paid for this place he also said, 'And I expect to
make at least two thousand per cent profit—if I don't, I'll be
disappointed'. Myself, I believe he meant it.

"Well, there you are—you've had to suffer a quarter of an
hour of bad-mannered slander from a self-confessed
scrounger. Put it down to my stomach if you feel charitable,
and my nature if you don't—I shan't mind. And don't forget
you invited me to unburden myself."

VII

I WONDER now that I didn't foresee the turn events might
take, or at any rate that the possibility failed to cross my
mind. It will have been obvious to you for some time, I dare
say, and therefore you will assume that it ought to have
been to me. All I can say is that if I was slow-witted then, I
am at least being honest about it now: it would be easy
enough to pretend that I practically started looking for the
ruby before it disappeared.

It was Hinkson who gave me the news the next afternoon, Wednesday. He had been talking to Lina in the
common sitting-room for over half an hour. Then he came
out with a very worried expression, saw me opposite in the
room for men only, and came straight across.

As he did so I noticed, not for the first time, that his
dark-blue suit was remarkably well cut, and plainly not long
out of the tailor's hands. I have seen dukes' sons in worse
when they were trying to look smart.

"Can I talk to you for a few minutes, Mr. Purdon?" he
asked, without preamble.

"Of course" I said. "Anything the matter?"

"Well, yes,"—as he sat down, carefully preserving the
creases in his trousers. "In fact, very much so. Can I rely on
you not to repeat anything I say?"

"Provided I'm allowed to stop you if I don't think I want
to hear it."

He smiled for a second at that, and then became more
serious than I had ever known him to be when he wasn't
playing up to Adair.

"I hardly think you'll have time," he told me. "The
matter's very simple. You remember the ruby in the box
with the parchment? Well, either Lina's lost the thing, or it's
been stolen."

For a moment I said nothing—too many thoughts were
crowding my mind. I don't know if anybody's ever timed
thought in terms of mental images per second, but I reckon
I must have come pretty near the speed limit then. The
pictures in question were mainly of five incidents: Montague telling me he was hard up; Montague on a different
occasion saying that the stone was worth at least four
thousand pounds; Buck beside me in the dark expecting
trouble; Tilly exposing the paucity of her allowance; and
Adair saying to Lina 'Look after it for me, Ladybird'.

"That's rather trying," I remarked at length, "Can I ask
questions?"

"Yes, please do. We're hoping you may be able to
help—that's why you're being told first."

"Really? I wouldn't bank on it if I were you, but I'll do
what I can, naturally."

"Thanks. You see, it's absolutely essential that the ruby's
found before the Major gets to hear. We don't somehow
think he'd care about having it lost."

"Not if I know him" I agreed. "Well, when does Lina remember seeing it last? Literally, I mean."

"Yesterday morning: she transferred it from her jewel box to her handbag."

"Why?"

"Because it occurred to her that she didn't really know if the two girls who come in every day are honest or not. If they got to know about the thing, one of them might perhaps be tempted."

"I see: so she put it in her handbag. The one she was using yesterday—the crocodile affair?"

"Yes: but—carelessly—she left it lying about between breakfast and the time she went out. Between half past eight and a few minutes after ten o'clock, say."

"Lying where?"

"On the dining-room table, next to where she'd been sitting. But the queer thing is that the bag wasn't there when she went to look for it. She found it on a table in the corridor near her bedroom, and it appears that Tilly put it there. She noticed it in the dining-room, she says, carried it upstairs to give to Lina, knocked at her door but got no answer, didn't like to go in, and so put it down where it was bound to be seen."

"Not dreaming that the ruby was inside, of course?"

"Oh no, naturally not."

But his answer lacked enthusiasm, I thought.

"Now," I said, "let's be a little more precise. Where-abouts in the handbag was it?"

"In the centre compartment, a sort of small built-in purse lined with chammy leather and opening with one of those ball-bearing gadgets."

"Yes, I know. Was there anything else in that com-partment?"

"She says not."

"And I can assume that anybody opening the bag casually and looking inside wouldn't have seen the stone?"

"No."

"What did she usually keep there?"

"In the centre purse? Money—at least, paper money. Small change she carries in the outer part, where it's easier to get at. You know, you're asking me almost exactly the same questions I asked her."

"Splendid—then I shan't feel such an oaf when I get stuck presently. When did she discover that the ruby was missing?"

"Yesterday afternoon while she was with you in Gloucester."

"Really? Perhaps that accounts for her silence: on the way home. I hope she isn't thinking I took it?"

Hinkson assured me earnestly that she wasn't.

"My dear man, of course not—or you'd be the absolutely last person we'd approach for help. It's because you seem to be one of the few people who couldn't have had anything to do with it that you're being told."

"Why couldn't I have done?" I asked.

"Well, your bedroom's nowhere near hers—you don't even use the same stair-case. You left the dining-room after breakfast before she did, and went out into the garden. When you came in you didn't stop, but merely collected your hat and coat from the hall and made for the garage: I saw you myself."

He paused, almost apologetically.

"I hope you aren't minding being told just what you did like this. It does rather sound as if we suspected you, but we had to rule you out definitely before saying anything."

"Oh, that's all right. And Lina's sure I didn't have an opportunity on the run, or in Gloucester itself?"

"Quite sure: even if you produced it now, we shouldn't believe you."

As he said this he smiled with more than frankness, as one honest fellow to another, and then became sober again, waiting for me to go on with my questioning.

"Very well," I agreed: "then as far as I can see, the position's this. If the ruby was lost, it may have happened any time between when it was put into the bag before breakfast yesterday morning and when its loss was discovered, and anywhere between here and Gloucester—more probably there than here, I should think. On the other hand, if it was stolen, then it was almost certainly stolen in the house."

He stopped me at that.

"But why?" he asked. "It'd be nice to be sure, but can we?"

"Yes, if what you've told me is correct. You see, you said that Lina always carries paper money in the compartment where she put the ruby, and only paper money. Well, you probably know that I lent her a fiver yesterday because she'd come out without anything. I actually gave it to her just before we separated for an hour before lunch, so I don't know what she did with it, but I assume she put it in her bag, and that it was then she discovered the ruby was gone. Is that right?"

Hinkson regarded me with obvious—too obvious—respect.

"I say, you're a detective yourself!" he exclaimed: a remark which I didn't fully appreciate till later. "Yes, you are perfectly right—only of course Lina didn't say anything to you at the time because she wanted to be absolutely certain she hadn't left it behind here,"

"Why, was there any room for doubt?"

"Well, no, not really, only you know how you'll believe anything rather than the worst. Yes, that does seem to make it fairly definite that if the thing was stolen, then it happened here: or if it was lost, for that matter. She says she didn't part with her bag after she left the house till she found out, and didn't open the middle purse between the time she put the ruby in and the time she went to dispose of your fiver."

"In fact, the thing to assume is that somebody in this house has pinched it."

He nodded thoughtfully.

"I'm afraid so," he said. "It's the most hopeful feature of the whole business, really. We do seem to stand some chance of getting it back before Adair hears, but if it had fallen out heaven knows when or where, or been taken by a cloak-room attendant or somebody like that, we wouldn't stand an earthly."

"No, that's sense. Very well, let's go by probability. The ruby was stolen in this house, and therefore by one or more of Tilly, Mr. Montague, Andrews, Buck, the cook, or the two village maidens."

"Oh, you can count them out. They didn't arrive yesterday till Lina was starting off—she met them half way down the drive."

"And what about the Judds?"

"I don't think so. He didn't come into the house before 11.00, and though Lina did drop in to see Mrs. Judd, she didn't put her bag down, she says."

I thought privately that 'drop in' wasn't a very suitable way of describing the visit in question, but said nothing. It was just possible that the story the girl had told me was true, I supposed.

"And similarly," I went on, "you don't suspect the Major, Lina, or yourself?"

He took this for a witticism, agreeing that there were five persons to consider.

"But Mr. Montague was in bed," I objected: and Hinkson's face became utterly grave. He glanced at me a moment significantly with his clear brown eyes, and then answered the point in a level expressionless tone.

"His bedroom is next door but one to Lina's on the same side of the passage," he informed me. "Between them is a bathroom, so-called." (I think I have already hinted that drainage wasn't Mauberley Grange's strong suit.)

I looked at him as steadily at this. His face was well formed, with a high intelligent forehead, a straight longish nose, and lips inclined rather to prominence. His chin stood out firmly, and altogether he was a personable young man: though perhaps not so young, at that, since the hair above his ears was lightly dusted with grey. He couldn't have been more than thirty-five, though, and the grey was evident only if you studied him very closely. He stood my scrutiny well, and had an answer ready for my next question.

"Is that idea to be considered seriously?" I asked.

"The loss of the ruby is a serious matter," he returned promptly. "I can't help feeling that we ought to consider *all* the possibilities."

"Yes, I expect you're right: he could have had an opportunity for taking it, as far as anybody knows. And Tilly had one too, of course, in that she handled the bag. But then, she admits to that openly, doesn't she?"

"Yes, true: but she's very short-sighted, you know, and she mightn't feel confident that she hadn't been seen carrying it upstairs."

"And why would she carry it upstairs?" I asked. "Why not leave it in the dining-room?"

"Oh, I don't know—she might think it would be more convincing if she didn't seem afraid to be connected with the bag. Please don't think I'm trying to incriminate her or anything, because I'm not: all I'm concerned about is getting the ruby back. Then there's Andrews: he says he doesn't remember seeing the bag downstairs at all, but I wasn't quite satisfied with his manner. I may have been fancying things, though—I'm afraid I haven't much experience of this sort of business."

"Downstairs," I repeated. "Does that mean he remembers seeing it upstairs?"

"Yes—on the table I mentioned, when he went to collect Mr. Montague's breakfast tray."

"I see. And the cook?"

Hinkson permitted himself a delicate shrug. "I hardly think she's worth bothering about, really. For one thing, she wouldn't have any reason for coming into our part of the house. I certainly never remember seeing her anywhere but in the kitchen or round by the refuse bin, do you?"

"No: and she's very stout, and couldn't possibly make a quick get-away. I'm inclined to count her out, too, for the time being. That leaves Buck: anything there?"

"I don't know," said Hinkson slowly. "Nothing that I can point to definitely, anyway. As a matter of fact, I've been rather chary of asking him questions. He always strikes me as being exceptionally alert, and I don't know him well enough to trust his powers of discretion."

(Yet he managed to say this in a way which conveyed the impression that he markedly distrusted them.)

"It's possible that he went into the dining-room while the bag was still there," he continued, "and if so that might easily account for everything. I'm pretty sure he wouldn't have gone upstairs on Lina's side of the house, though—he sleeps your way."

"Really? When?"

"Ah, when? He does seem to do with remarkably little shut-eye, doesn't he? I don't want to wrong the man, but I don't somehow think he'd get on very well as a sentry in the army.

"Well, there you have the whole position, Mr. Purdon. The ruby has definitely gone, theft seems more probable than loss, and theft almost certainly implicates someone here. On the face of things as I see them, I'd say that the most likely culprits are Mr. Montague, Tilly, Andrews, or Buck, in approximately that order. May I repeat, most likely on the face of things? I've no grudge against any of those four persons, and I don't think Lina has either. If one of them should turn out to be a thief, I'd feel I knew less of human nature than I imagined I did. All the same, you can never be sure of people till you've seen them buried, and even then one in every few million taps on the inside of his coffin and asks for water. Much as I dislike coming to that conclusion about the probable identity of the probable thief, I feel it's the most sensible one. If you can offer any sort of help, either in the way of alternative theories, or of ways and means for tackling the detective part of the business, we'll be extremely grateful."

"Well, I'm willing to do what I can," I said: "only what can I do? I'm no sleuth, and to go round and ask the suspects a series of pointed and exhaustive questions is about the last procedure, as I understand the situation. You might just as well send for the police and have done with it. If Mr. Montague's lying on the ruby tight under his mattress, for instance, he obviously isn't going to roll over obligingly and let anyone pull it out. What I imagine he'd do—or the rest of them, except perhaps Tilly—would be to get in touch with Adair and complain of the way he was being treated. It isn't an easy problem at all, and I'm hanged if I know why you came to me for help."

Once more Hinkson looked at me straight in the eye; only this time contemplatively, as though he were summing me up before revealing even greater secrets.

"It's certainly not easy," he agreed. "It's about the hardest thing I've ever struck in the way of a problem, not forgetting the job we had to find the wretched thing in the first place.

"Look here, Mr. Purdon: I'm going to be honest with you. I came to you for three reasons. The first I've already mentioned—because you at least were clear of suspicion. Secondly, I felt the need for telling someone: it's an unpleasant kind of burden to carry about, and anyway, if there were something practical that we could do, it would have to be undertaken by somebody other than me. I shall be cooped up in the study 'helping' the Major—I ought to have been there ages ago. Thirdly, and most pertinent of all, I wanted *to* find out if you'd *really* do something for us.

"To put things plainly, I want to ask a considerable favour, which you'll be perfectly justified in refusing. After all, the ruby's nothing to you one way or the other: nor is the question of how Adair might take its disappearance. Incidentally, the reason I'm doing all the talking instead of Lina

is that she fancies, rightly or wrongly, that you've met
people you like better than you like her, and she felt you
might automatically shy at anything she suggested.

"Mr. Purdon, I heard you say once that your best friend is
an Inspector at Scotland Yard. Would it be at all possible for
him to come down here to look into the business unofficially?
No, let me explain what I mean,"—as I made signs of being
about to expostulate. "I know it's an outrageous thing to
suggest, but desperate ills require desperate remedies, as
they say, and it does seem to me to offer the very best
chance of ever seeing the stone again.

"Briefly, this is the idea. You see Adair *before* you get in
touch with your friend, and say—if you don't object to
telling deliberate lies—that you hear he's got a few days
with nothing particular to do, and would very much like to
come down for a breath of fresh air: may he do so? If Adair
says no, of course, that's the end of the matter: but there's
a fair hope that he wouldn't, I think. You then ring up your
friend, put the position before him, and find out if he'd
consider taking a week off to do what he can.

"Naturally, it may so happen that he couldn't do that
even if he were willing; but, again, perhaps he could. If he's
agreeable, you arrange for him to come down as quickly as
he knows how. If he isn't, you apologize on my behalf for
troubling him, and tell Adair your friend's very sorry, but he
can't get away after all.

"The one thing that bothers me is the question of, well,
of recompense. If I asked a solicitor friend of a friend of
mine to do a job of work for me, I'd expect to pay him his
usual fees. In this case it'll be a bit different, because
anything your friend might be able to do if he came would
have to be unofficial. If you think that difficulty could be
overcome, and you yourself don't veto the plan at the
outset, then there may be a hope of repairing the damage.
Otherwise, I'm pretty pessimistic. Anyway, thank you for
listening so patiently. What do you say?"

If I had put it into words, it would have been that in some
unaccountable way my opinion of Hinkson had gone up.
The sole reason I could see for his interesting himself in the
affair so greatly was that he must be genuinely fond of Lina,
and while that alone would make me think well of no man,
yet there was something in his manner of putting forward
his preposterous suggestion which I couldn't help liking. He

wasn't servile: but nor was he high-handed. If I would help him, he seemed to be saying, by getting Beale to help, he would be extremely grateful, because in so doing I should also be helping someone he cared about. If I wouldn't, then my refusal would be understandable, although a matter for regret.

To begin with, of course, I had found his plan for re-covering the ruby utterly ridiculous. As far as I could see, Beale would be no better placed than ourselves if his hands were tied by the removal of all his authority. People would resent prying questions from him quite as much as from us, and probably more so. Yet the project had two points in its favour. It would give Beale a short holiday, which I was sure he could do with, and it would enable me to verify my own conclusions about the atmosphere at Mauberley Grange. You will remember that I had already suggested that I should try to get him an invitation to join us: to which he had made no response.

As Roger Montague had said, we were a rum lot. I was certain of that, yet I felt that I would greatly like Beale's own opinion on the matter. I value his judgment of men more than that of anyone I know, and I thought that on his side he might derive some interest from meeting Buck, if no one else. Buck, I had decided, with his perpetual air of pretending that he wasn't really as smart as he knew he really was, with his aura of competence and his practical abilities in the way of shooting and night-stalking, ought certainly to please my friend.

"I don't think it's the least bit likely that he'll be able to come," I said now, as Hinkson waited attentively, "but I don't mind ringing up to find out. It would have to be on his own terms, though."

"Of course, of course. I say, it's thundering good of you, and I'm sure Lina will be delighted. You'll see the Major first, though, won't you? And where will you telephone from? Not the study for preference, I should say."

"No, I'll go out exploring," I told him: "providing Adair doesn't object to an extra visitor. I don't see why he should, though, do you?"

"No: at least, I sincerely hope not. By the way, is your pal a big fellow with fair wavy hair and glasses? If so, I've an idea I may have met him."

I shook my head.

"He isn't particularly big, his hair isn't wavy, and he certainly doesn't wear glasses," I said.

"In fact, it's not the same chap, what?"—and Hinkson laughed as heartily as if he had made an especially witty remark. Win or lose, his manner seemed to say, I was his friend for life.

VIII

I WENT along immediately to the study and tapped on the unpanelled door: which, by the sound produced, wasn't nearly so solid as I should have thought. It couldn't have been more than an inch thick. I heard then the unexpected rattle of a bolt, the turn of the key in the lock, and Adair was looking at me through a narrow crack.

"Oh, it's you," he said, not ungraciously. "Come in."

I did so, passed some light remark about the door's fragility, and was given a short lecture on the sins of Victorian domestic architecture.

"Some purblind fool started meddling with this place around 1850 or thereabouts, at a guess," declared the Major. "Luckily he didn't get very far—died of a bad conscience, perhaps, or somebody with a ha'p'orth of taste threw a chandelier at him. This is the room he did most damage in—look at the hideous fireplace, and that revolting mantelpiece supported by ill-proportioned infants. I ask you! Marble and old oak together! The door is obviously his handiwork, or that of the inefficient cheap-jack he employed. It's a rotten bad fit, anyway—lets in a draught like a monsoon. There are one or two other equally regrettable mistakes tucked away in odd corners, but this room is easily the worst."

Adair's change of manner surprised me a little. For the last few days he had been very reserved and grumpy, all his attention centred on his problem: yet here he was, suddenly, almost back to his old self, inclined to be chatty, never reluctant to display any knowledge he possessed which his audience didn't, and always with his own peculiar air of distinction about him.

Picture the man now, sitting at the round swivel chair before his flat-topped desk, dressed in biscuit-coloured plus fours and an Army tie, his white hair smoothly clinging to

his square head, his short grey moustache wholly sym-
metrical, and his deep-set green eyes as clear as a
youngster's, though he could have had scarcely any exer-
cise in the last week. He is alive, and fit, and good for
twenty years, you'd think.

I looked round the room, barely furnished with a small
table, two chairs, a massive book-case, and a cheap plain
carpet as well as the articles I have mentioned. The contents of the book-case overflowed in all directions: there
were toppling piles of volumes on the floor and the wooden
seat lining the bay window, most of them open and none by
the appearance of them new. I saw too that the bolts on the
door into the corridor were recent additions, and voiced my
curiosity.

"Oh, just precautions," said the Major airily. "When I'm
working I like to feel that nobody can possibly come in
without my knowledge."

"And why do you use this room anyway?" I pursued.
"Why pick the only one with a marble mantelpiece?"

In reply he pointed to the leaded window-panes, beyond
which was a small lawn freshly rolled, and two or three
beeches. Farther back stood a hedge of holly bright with
berries, and clumps of rhododendrons which promised to
be worth seeing later.

"At the moment that's the only bit of all the acres and
acres of gardens that I can bear to look at," he told me,
seriously, as if his aesthetic sense were abnormally developed and guided all his actions. "When we get settled in
more I may change, but I'm hanged if I want to be faced
with a tangle of weeds, and flowerbeds that are worse than
the abomination of desolation. That's supposed to have
been a heathen statue, though, isn't it? Or don't you know?
But any time you're short of a job you might try your hand
at an odd corner of the garden—it'll all help. And put a bit
more on the fire, will you? Rake it out from the bottom
first—the coal's absolute filth, and all the small logs are
damp."

I smiled as I obeyed his bidding. This was the Adair I
knew, and thought once I had liked: the man who looked as
if he could outrun you at any distance up to three miles, as
if he needed but the slightest excuse to become enormously energetic in physical action; and who invariably
never raised his own finger if someone else would do it for

him. It may have been a pose, of course—I was never certain.

"And how's the labour going?" I asked presently. There was no hurry about coming to the point, and I was sure that sooner or later he would query the reason for my visit.

"Well, one mustn't grumble, Tony. I'd be the last man to boast about what I haven't done yet, but at least I've made a start on the thing."

"Really? That's good news. You've kept very quiet about it, though."

"Oh, I didn't see any sense in raising false hopes. I may strike a snag yet—not that I expect to, I must say. No: in confidence, I fancy I've got old Jasper's measure, and it's just a matter of steady slogging."

"Well, that's more cheerful. What put you on the right track?"

"Experience, as I told you beforehand. It'll all be so much Greek to you, but I don't mind explaining. I happened to have made a pretty thorough study of shorthand and its history: it always pays to tackle a subject properly if you're going to tackle it at all. The first real system was brought out in 1602 by a chap named Willis—John Willis. *The Art of Stenography* he called it—literally 'writing narrowly'. And after him came the flood. Between 1602 and 1837, when Pitman introduced his phonography, there were about two hundred different shorthand systems made public, and nobody knows how many private ones.

"Now, I wasn't concerned with anything written after 1695, the date on the box, and there were two possibilities before me. Either Mauberley's shorthand was entirely his own work, or it was an adaptation of an existing system: I could see at a glance that it wasn't one that had ever been published. The second possibility was far more likely than the first, so I began to look at things from that point of view. Now, the best shorthand author in the seventeenth century was William Mason. He made a start in 1672 with his *Pen pluck'd from an Eagle's Wing,* a system based on an earlier one by a man named Rich: one, by the way, which was still being used as late as 1847, which shows it wasn't too dusty. It contained over three hundred separate signs, though, and Mason's even more—a far cry from the simple methods of today.

"Well, Mason issued a revision of his system in 1682 called *Art's Advancement,* in which he retained only six of Rich's letters. He also wrote a third book in 1707, *Plume Volante,* but that doesn't affect us, naturally. Now, as I said a moment ago, I felt that if Jasper Mauberley was really interested in shorthand, as he must have been, he'd take care to be well up in current knowledge, so I quite expected to find traces in the clue of Rich or Mason, or maybe both. The trouble was that I had three systems to choose from, and plumped—wrongly—for the latest one, *Art's Advancement.* It turned out eventually that the stuff on the parchment has a much greater resemblance to the original 1646 Rich, with a dash of Mason, 1672 brand, and a good intermixture of his own ideas.

"Well, there you are—and I don't suppose you're an atom the wiser. Still, I am, and that's what counts, isn't it?"

And at that he stretched his arms above his head suddenly, and smiled, and I could see that in reality he was extremely pleased with himself.

"Do I look tired?" he queried, a moment later.

"Yes, a bit, perhaps."

"And no wonder. It was three o'clock before I turned in last night, and quarter to five the night before: and up again at 8.00 sharp. But no more of that: from now on I'm going to be patient. Seven hours a day—no more, no less."

"And how many days?"

"Two more, with luck: I reckon to finish on Friday evening. At least, always supposing today *is* Wednesday?"

"It is," I said. "How long will the final version be?"

"Rather more than I estimated. A thousand words, about—say three to four pages of an ordinary novel."

"And a sight more interesting. Does it make sense as far as you've gone?"

"Pretty well."

He shuffled through the mass of papers on his desk for a moment, chose one, and began to read.

" 'Herein do I, Jasper Mauberley, Esquire, being already an old man and without great expectation of life, and cursed almost from birth with a deformity of my body, purpose to assist him that hath the wit to read my meaning to such delight as gold may purchase or the rarest gems afford. Albeit, the wit must needs be there first, and in full measure, and a little learning also, and a great diligence of application, and withal a cunning apprehension. . .' And so on. He's apt to be prosy, and I shan't be at all surprised to find that the actual directions don't come till the last paragraph or so. One daren't take a chance, though. Jasper was a wily bird, and quite capable of dealing appropriately with impatience. 'Straight through where possible' is the cryptographer's golden rule, or one of them."

He paused, nodded solemnly, and began to fill his pipe, having apparently forgotten that I was his cigarette provider.

"Don't go telling the others anything," he said. "Some people's voices carry a long way, and there may be ears waiting to pick up an odd word. I don't know why I've talked so much to you, really: because you look honest, perhaps."

"Thank you. If so, it was the only gift the fairies gave me in the way of looks."

"And a good one. Well, I must get back to work—unless there was anything you wanted to see me for?"

His way of putting the question was characteristic of the man. He certainly did have patience, underneath his occasional brusqueness and bullying. At least, perhaps I had better qualify that. He could be patient over things, or people, if he was interested in them; otherwise the rough side of his tongue was good enough, as for Tilly. Why he should ever have taken the slightest fancy to me I honestly can't say: but so it was, or I wouldn't have been in the house at all. He made his request for information now in an ordinary voice, but I felt certain from the attentive glance which accompanied it that for some time he had been wondering about the reason for my visit: yet he raised the point as a mere apparent afterthought.

"Yes, there was something," I agreed. "You may think it awful cheek, but I'm going to ask you to let me invite a friend down here for a few days. I think you've heard me speak of him—Chief-Inspector Beale of New Scotland Yard, If you remember, I got your permission to tell him what I was doing a week or two back."

"Scotland Yard, eh?" murmured Adair, his bushy eyebrows lifted, and he regarded me quizzically. "Coming on business?

"Good lord no! I'd better make the position clear. He's taking a week off that's been due to him for months, and wants to know if I'll join him for a prolonged pub crawl. Well,

I don't want to let him down, but on the other hand I emphatically don't want to miss being in at the death here, so I thought maybe you wouldn't mind if we amalgamated things. I think you wouldn't dislike him: he isn't at all the big-booted bowler-hatted sort of policeman, and I promise he won't get in the way or be a nuisance. And as well—don't think me impertinent, Major—what you've just told me about expecting to finish the clue by Friday, and all these bolts and things, makes me feel that it mightn't be a bad idea to have somebody in the house who'd know what to do if there was any funny business."

At this he frowned a little.

"Funny business?"

"Yes—or have you dropped the Warner theory?"

"Oh, I see. I don't know—we haven't seen anything of him. Still, there may be something in what you say, and I can't see any objection to your friend's coming down if he's content to rough it."

He stopped for a moment, thinking.

"He'd have to sleep in your room," he said, "and you'll have to go out and buy him a bed or a hammock or something—we've no spare kit here, you know. If that's all right, then fix things how you like. Want to use the 'phone?"

I sought quickly for an answer. Adair wasn't the kind of man to go out of the room while I spoke to Beale, and I didn't see how I could deceive him and yet make the position understandable the other end.

"No, I don't think so, thanks—he doesn't care about being rung up during the day, and anyway it's ten to one he won't be available. I'll write."

"As you please. And when you do you might ask him to bring down a copy of the *Barrack-Room Ballads,* will you? I've had a couple of lines running through my head for days, and I can't be certain if I've got them right, and I hate not knowing."

"Yes, I'll remember: and many thanks for letting me invite him."

"Not at all. If you see Hinkson send him along, will you? And not a word to anyone about how I'm getting on, mind."

I talked to Beale for twenty minutes on the telephone that afternoon from a call-box in the next village but one. He told me that I was lucky to get hold of him, as he had

only just returned from a two-day trip to Bedford in connection with a jewel robbery.

"Splendid," I said. "Jewel robberies must be fashionable
just now—we've had one here." I then acquainted him with
the details, and with Hinkson's suggestion that he should
come down to Mauberley Grange.

"Yes, I know it's a daft idea," I agreed, after he had said
what occurred to him about it. "I know you're not at everybody's beck and call, and can't lift a finger outside the
home counties without some chief-constable's gracious
sanction, and all the rest of it. All the same, I rather wish
you'd come, Ted: not so much to find the ruby as to have a
look at the household, and tell me what you think. After all,
I did feel that something was wrong, or about to go wrong,
and it has. And you did say you were interested."

"I know—I am: but what in the world could I do if I did
come?"

"Keep me company," I said. "There's precious little of it
here that I'm frantic about, and yet I don't feel like
chucking my hand in when the treasure's all but found. I
told Hinkson that if you did turn up it would be on your own
terms, so you can always look mysterious and deep in
thought when he's about, and say you're considering the
matter from another angle. *Could* you get away?"

"Yes, I dare say I could," he admitted. "Who do you think
took the thing?"

"I don't know. I wouldn't put it past any of them, including Lina and Hinkson themselves. But probably not, on
second thoughts, or they'd hardly be so anxious to have
you down, would they?"

"Why not? Attack is the best form of defence, and all the
rest of the copybook. Criminals—amateur ones—are mostly
pretty vain, as I've told you repeatedly. Your Lina and
Hinkson might feel confident of their ability to diddle all the
detectives from Dupin to Fortune, and to invite investigation is always a sure draw with simple-minded people like
you. Still, the telephone's no place for a lecture, even
though you're paying for the call. Shall I come or shan't I?
I mean, shall I try to come? What's the country like?"

"Glorious," I told him. "Roses in bud, honeysuckle
smothering the strawberries, and green peas all day every
day."

"I see: and as much of the treasure as I can carry away in a large bucket, I suppose. Well, I'll see what I can do, only don't be too optimistic to Hinkson. Tell him that the most I could do would be to act in a strictly advisory capacity, and that I shan't bind myself to go any farther than speculate in the most comfortable chair, or in bed if it's really cold. I'll drop you a line tonight—can't find out now because my superiors are in conference, and mightn't take it kindly if I barged in to ask for a week off. It won't be till Friday, though, anyway, because I'm going to Queen's Hall tomorrow. And I can't promise to stay longer than Sunday evening if I do come: but anybody's entitled to a long week-end now and again. Where shall I book to?"

"Oh, Gloucester—I'll pick you up there. Oh, and get a copy of *Barrack-Room Ballads,* will you? Adair wants one. I'm glad you haven't spat on the idea—I was rather afraid you might."

"Yes? Mustn't spit into telephones—t'ain't hygienic. And I'm thinking of myself quite as much as of you, anyway. There's a case of fraud in the offing, all very dull and complicated, and involving hours of tedious interviews and ledger-hunting, and maybe if I can get out of town on Friday I'll escape it."

IX

THE next day there was a letter from Beale to say that he would arrive at Gloucester at 9.30 on Friday evening. I duly told Hinkson, who appeared delighted and thanked me profusely, partly in my own right and partly as Beale's proxy.

"And I only hope Adair doesn't get awkward before then," he added. "He didn't mention the ruby to you? Good."

However, that paragraph shouldn't have come next, strictly speaking. On the same evening, Wednesday, an incident took place which gave us all much food for thought.

Andrews, the butler, went down about ten o'clock to post a letter in the box at the bottom of the drive, just outside the wrought-iron gates. He had remembered—he said afterwards—that Friday was his sister's birthday. She lived in Ireland, and he wanted his gift, a ten-shilling note, to reach her the right day. Posted on Wednesday night the letter would start its travels at 6.05 the next morning, instead of at 10.25, and one post might make all the difference, he thought.

He went down the drive, I say, a few minutes after ten o'clock; and a few minutes after that, at something short of quarter past, Hinkson departed on a similar errand. He had been stewing in the study almost ever since I sent him along after my chat with Adair and—as he put it—he was about fed up with the stench of his employer's tobacco smoke. He told me that himself just before he set out—I was doing the *Telegraph* cross-word when he looked in to see if I had anything to go.

Exactly what happened can be briefly summarized from his own story later. He was running down the drive in the starlight when he heard a shout some distance away, and a scuffle. That made him run even faster, but he could see nothing, though he made use of his torch. Then a second cry determined that whatever the commotion, it was taking place among the bushes to his left. He stopped to listen, and thought he heard the sound of someone hurrying away. He called out himself, but there was no answer, and it took him five minutes to find Andrews lying half dazed in a laurel, his nose bleeding profusely and his right eye closing rapidly.

The letters went unposted that night. Somehow he helped the butler back to the house, and very soon we were all pressing round, plying the victim with whisky and demanding to know what had happened. Even Adair was fetched from his labours, and of course took the lead in the interrogation.

Andrews wasn't able to tell a very coherent story, though. All he remembered was hearing a strange sound as he made his way to the post-box—a noise to his left that might have been a clumsy footstep crackling a dead branch. He turned, could see nothing because the battery of his torch had run down, nevertheless went off the path a yard or two, and was at once attacked. He did what he could to defend himself, but that was little enough, he said, and none of us disbelieved him. He seemed the last man to give any of kind of account of himself in a hand-to-hand fight—he was too feeble-spirited and placid. What his attacker had looked like he could scarcely make a guess at: a big man, he thought, from the weight behind his blows, and dressed in something rough and dark.

His injuries weren't so serious as we had at first imagined. Once the bleeding from his nose had stopped, and as much as possible of the consequent mess cleared up, he was left with a splitting headache and a blackening eye. It seemed, however, that he had been hit with nothing worse than a fist. Adair was very much upset, and eagerly assented to my suggestion that I and Hinkson should turn out Judd and make a thorough search of the grounds. Very solemnly he went along to his study, and returned with one loaded automatic pistol, one ornamental sword-stick too fragile to be much use except to an expert fencer, and one shillelagh that I had seen hanging on the wall by the book-case. He handed the pistol to me, and I hurriedly attempted to pass it on to Hinkson, saying that I wasn't to be trusted with anything that hadn't a butt and two barrels. That didn't work, though, because the secretary suddenly surprised everyone by declaring that he already had a gun of his own, and would fetch it; and accordingly I spent the next two hours in constant dread of tripping up and despatching myself ingloriously to wherever I am bound for when I do depart. Not till I had returned the accursed thing to its owner did he bother to show me how it worked, and assure me that without releasing the safety-catch I couldn't have done anything with it except as a club.

For all our efforts, and we really took the business seriously, we found nothing but a little blood at the scene of the struggle, and rough indications of the line of flight taken by the unknown aggressor. Of anything remotely like a clue to his identity we were disappointed: the ground was not in the right condition to take clear footprints. On our rounds we ran into Buck, who had heard nothing amiss, informed him of events, and got him to join us. We then split up into pairs, to cover the woods more quickly and stand a better chance of finding a possible fugitive. I had Judd for my companion, and though I shouldn't have picked him for choice, I must say he was better than nobody. He seemed to think it a great shame that Andrews should have been the recipient of any violence that was abroad, and made it fairly clear how much he would have preferred to hear of Hinkson's victimization.

"And who could it have bin, anyway?" he asked me. "That there Warner the guv'nor sacked for stealing? Garn—he hadn't got the strength to 'it a dead rabbit. A proper wet fish 'e was: couldn't bear for me to drive him at more than thirty-five, 'cos it made 'im nervous. 'Judd' he'd say,' 'it's me 'eart—I 'd a bad turn when I was a nipper and I ain't strong'."

"But maybe he was having you on?" I suggested.

"Eh?"

"Pretending to be all limp and flabby so that nobody would ever suspect him of violence. Part of a deep-laid plan, of course."

"Garn, don't you go kidding me, Mr. Purdon. He was like a wet leather when he come, and what'd 'e got to pretend about then? Why, the guv'nor hadn't even started this 'ere treasure 'unting."

"No, that's true. Tell me, what did the fellow steal? I've heard so much about the mysterious theft for which he was sacked, but not a word about what it was he pinched."

"Books," said Judd, with manifest scorn. "Just 'alf a dozen books! Signed or something I think they was—any old how, the guv'nor fair raved about it. Anybody'd have thought it was a barrer-ful of gold, the way 'e carried on; but books—I ask yer!"

It will be seen that Judd had little appreciation of the written word; and also, I hope, that he was by no means lacking in shrewdness.

Another point about Warner occurred to me as we trudged through the lonely woods, peering continually into the darker shadows, and now and then pausing to listen intently. If the man had had designs on the treasure, perhaps enticed into belief in its existence by Adair's enthusiasm, would he have been silly enough to get himself dismissed for the sake of a few signed first editions? (I took that to be Judd's meaning, and later verified it from Tilly.) Wouldn't he have hung on at all costs until the treasure was at least vaguely located? If his dismissal had been unintended, then I thought he must be too careless a conspirator to be much afraid of: if it had been deliberate, then I utterly failed to see the reason for such a maneuver. However, I silently conceded that there might still have been a point: and somebody had certainly hit Andrews very hard.

By the next morning, Thursday, his eye was closed, and most unhealthy in colour. On my own initiative I ran him in to see the nearest doctor: nobody else seemed to think of arranging for him to have attention. He expressed gratitude, and came out of the surgery with a bottle of lotion and instructions to apply it on a pad at frequent intervals. He then somewhat timidly asked me if I thought an eye-shade would make his appearance less unsightly. I agreed that it easily might, assured him that Mauberley Grange could very well do without his services for a couple of hours, and ran him into Cheltenham. I wanted to telephone Beale, to tell him the news, but was unable to get in touch with him; and I also wanted to purchase a camp bed and suitable accessories for his stay.

The eye-shade we bought later was even more hideous than the bruised eye, in my view. It was of a kind of puce colour that looked like one of Lina's fingernails greatly enlarged: for which reason we, or rather I, chose it deliberately. I have a good memory for colours, and the moment I saw it I said, "Let's have that one, and cheer the place up a bit." I hoped Lina would be alert enough to take the hint, and I intended to make sure she knew the thing was my idea. The point proved to be non-existent when we got home, however: the wretch had changed her scheme of decorations again to pillar-box red, and I felt unreasonably annoyed.

And now, I think, I shall have to decide whether or not to be entirely honest about my feelings towards Lina. Up to the present I've been giving the impression—I hope—that I thoroughly disliked her: but that wasn't by any means the whole of the story. There are some people who look for no more in a woman than a high standard of physical perfection: let her have the face and figure of a film star, and for all they care she can combine these with the mind of a near-lunatic and the disposition of a vixen. There are, again, other people who maintain that the physical side of the matter deserves hardly any consideration at all: not even barely, if I may pun. At least, that's what they say, and some of them look as if they might be serious. Myself, I don't understand leaving the physical out of it altogether: after all, people are solid substances as well as being intellectual or moral entities and influences, and their solidity is what meets one's eye, and since matter necessitates form, I'm always in favour of beauty rather than ugliness, there being any choice. I might be capable of adoring a monstrosity if she was also a saint—I don't know, and rather doubt it; but I'm quite sure that I should be more capable of adoring a devil if she was attractive enough. If you say that there can be no personal beauty without goodness, I can only reply that the weight of recorded experience is against the view. Many of the famous beauties of history were not what anyone could call good: consider Circe, Cleopatra, Cressida, or Clytemnestra, to deal lightly with but one letter of the alphabet.

Now, if I could have forgotten all qualities but those of flesh I should have been well content with Lina: hers in that respect were high. Luckily—I think—I wasn't able to; or, more truthfully, I was able not to. The girl might not be exactly a devil, but I would have wagered body and bank-balance that she was no saint, and not even a saint's very distant cousin. It was because I wanted to run no risk of succumbing to her physical charms at the expense of better judgment and self-respect that I made not the slightest effort to overlook her bad points, or pretend they weren't really bad. She was lazy, insincere, vain, greedy, not devoid of cruelty, and very selfish—perhaps the last connotes all the others: and so long as I could remember these things I felt safe from any weakness in my own character. Perhaps it seems a queer way of behaving on my part: but then, I know myself better than you do.

For the moment, enough of her: I'll go back to Andrews and the eye-shade. He put it on in the shop, over a piece of lint soaked in his lotion, the assistant and I smirked at one another and lyingly remarked how well it suited him, and we went out into the piercing coldness of an east wind.

"Come and have a drink," I said. "A few minutes more won't make any difference, and I'll stand the racket if anyone tries to row you."

I made the suggestion because I rather wanted to get the man to talk, if possible. I'd been interested in him ever since the postman episode, but so far had had little opportunity for more than a passing word. He seemed to spend the days moving silently from place to place, always at hand when you wanted anything, always deferential, and always, I thought, a trifle stealthy.

He accepted my offer, but after three whiskies-and-soda he was as much a mystery as ever. If I attempted to extract information from him in a circuitous way he would answer as though he were being perfectly open, and could have nothing at all to hide: and yet he would say not a word that helped me. A sample of our conversation may make this clearer.

Self (being unpleasantly hearty): I bet that's the first time you've ever been jumped on in the dark and knocked about.

Andrews: Yes sir, and I hope it will be the last, I'm sure.

Self: So do I. And it seems to have been such a senseless attack. What could anybody hope to gain by laying you out?

Andrews: I can't imagine, sir. Possibly I was mistaken for someone else.

Self: Ah, that's an idea: but mistaken for whom, I wonder?

Andrews: I don't suppose we'll ever know that, sir.

Self (still being hearty): No, rather not. Still, we might try guessing. It couldn't have been Major Adair—he's much bigger than you are. Mr. Hinkson's about the only one I can think of—you're something of a size and build.

Andrews (doubtfully): But I haven't anything like his springy walk—though of course it wasn't at all light.

Self: No. You haven't any serious personal enemies?

Andrews: Oh no sir, I don't think so. I'm sure I haven't.

Self: And you wouldn't be likely to be carrying a roll of bank-notes or anything? I did know a man in your line once who always took his entire fortune about with him wherever he went—about seven hundred pounds it was, all neatly sewn into a special belt round his tummy. I should think he was exceptional, though. *(Story completely untrue.)*

Andrews: Yes sir, very probably. I had little of any value on my person, except the ten-shilling note in the letter I was carrying, and nobody could possibly know about that except Cook.

Self: And Cook couldn't run away, could she? At least, not from what I've seen of her. She might manage to re-cede slowly, but that's about all. Could the person who jumped on you have been Warner, do you think? Or weren't you with the Major in his day?

Andrews: Oh yes sir, some time before he was engaged. But I don't think it could have been Mr. Warner, somehow. He wasn't very strong, you know.

Self: So they tell me. And it couldn't have been Buck?

Andrews (with noticeable surprise): Buck, sir?

Self: Yes—it was an idea that occurred to me. Buck mistakes you for an intruder, piles into you, and then discovers his error. What does he do next? Mightn't he prefer to sneak off and say nothing rather than admit he'd made a fool of himself and a mess of you?

Andrews: Buck wears a raincoat at nights, sir, but I'm sure the man who attacked me had on something darker and thicker. I vaguely remember catching hold of cloth.

Self: I see. Then please don't tell Buck about my idea—he mightn't fancy it.

Andrews: No sir, of course not.

Self: He's a likeable little chap, don't you think?

Andrews: Yes sir, very good company, and a surprisingly clever mimic.

Self: Really?

Andrews: Yes sir: at times he borders on the disrespectful.

Self: Does he imitate the lot of us?

Andrews: Well sir, in a manner of speaking.

Self: I was never any good at that sort of thing—were you?

Andrews: No sir—I'm too self-conscious. I once played Father Christmas at a children's party when I was in service with the late Sir John Bywater, but the beard made all the difference, of course.

Self (giving up): Yes, of course. Have another drink.

X

I SPENT that afternoon—Thursday—following Adair's suggestion of making myself handy in the garden. Now that Beale was definitely coming I felt disinclined to do anything about the missing ruby myself. I mean, I could think of nothing sensible to do, and welcomed the idea of shoving the whole thing on to a real detective. Instead I dressed up in my oldest clothes, hunted out a rake and a small fork, and made a determined onslaught on one of the flowerbeds in the front of the house. It was clotted with thistles and burdock and bindweed, and the last proved so predominant and well established that its white stringy roots were still evident at a depth of two feet, as I discovered with the help of a spade. The soil was heavy clay, and my hands were soon in a mess. I continued only because I felt I ought to be doing something useful, or meant to be, if for no more than a couple of hours. Otherwise I should be too conspicuously appearing to follow the example of Montague and Lina: sitting round impatiently while Adair did all the work, in the hope that I should get a share in the proceeds.

Montague was up today, but not out of doors. He had settled himself comfortably in the ingle-nook of the men's sitting-room, and was alleged to be writing letters. I could just see his back from where I laboured unskillfully, but every time I glanced up he was plainly dozing.

Presently Lina put her head out of an upstairs window and called to me.

"Tony! Have you seen Tilly?"

"No," I shouted back, "not since lunch."

"That's funny," she said, in a lower voice. "I wonder where she's got to."

She then put her head inside again without attempting to explain why she wanted the missing girl: but I thought I had noticed an unusual gravity in her manner, as if for once it really mattered to her where Tilly was.

A few minutes later Judd came by, dressed in a leather coat and gum boots, and from the state he was in I surmised that he had been cleaning Adair's car, an old-fashioned but powerful Lancia which I wouldn't have minded for myself. He eyed me curiously, hesitated, and then came nearer.

"Ain't found nothing, 'ave you, sir?"

"Found anything? Only about a mile of this stuff,"—and I held up a dangling root for his inspection.

"Oh, that! Fair 'olds the ground together, don't it? I thought maybe you was digging for that there treasure."

"No fear—wish I were."

So Judd too was interested, I mused, as he grinned in agreement and moved on, swinging his arms and whistling shrilly. And yet, I asked myself, why not? He could, hardly help but be interested in something that engrossed everybody else's attention, and he was certainly justified in wondering at my unaccustomed activity.

Next I made a rough calculation, and came to the conclusion that at my present rate of progress it would take me the best part of two years to clear the beds and borders: by which time the place whereon I now squatted would be a wilderness again. The thought depressed me. I looked at my hands, I looked at the ground, I tugged irritably at a horse-radish root, conquered it—and raised my head at a slight sound to see Buck a few yards away. In his arms he was carrying Tilly's plump and damp and apparently lifeless body as easily as if she had been a small child.

I was by his side in a moment, sensing somehow just what had happened. At times the imagination is exceptionally alert to correlate disjointed facts into a coherent and convincing whole, and so it was with me then, though I claim no particular credit. I saw at a glance that Tilly's clothes were wringing wet, and Buck's too. She had on a tweed coat, from the edges of which water still dripped lazily, and her hair hung in limp hanks. However, she was obviously alive, at a nearer view, and that was something.

I wasted no time in asking questions.

"Not that way," I said, as Buck made to continue towards the arched porch. " 'Round the back—mustn't let anybody see if we can help it."

Fortunately we could get into the house near the kitchen without passing the study window. I kept my eyes open till we were safely indoors, and was pretty sure that we had been unobserved. Straight to the roaring kitchen range we carried her, bade the cook fetch restoratives and hot blankets and so on, which she did as quickly as her bulk would allow, clucking with dismay, while Buck and I worked our hardest to get Tilly back to consciousness. It was not a difficult task, and presently she was peering short-sightedly about: her spectacles had gone, and she could obviously see very little without them.

"Don't try to talk," I told her. "Just drink this, and stop worrying."

She obeyed the first instruction without protest, though usually, I knew, she hated the taste or smell of whisky. I must admit I was relieved to find her so life-like again, and more so when eventually she had been smuggled upstairs to her room unseen, and had solemnly promised to take off every stitch of clothing, towel herself vigorously, and get in between the cook's hot blankets. I said I would come back in an hour or so to find out how she was getting on, and that meanwhile she was to sleep, if possible, and anyway not to worry. Then I went downstairs, and out into the orchard with Buck, who had changed, and he told me his story.

He had been crossing the lower woods after fixing some barbed wire on the boundary of the estate: Adair had ordered half a dozen drums of the stuff by telephone early that morning from Gloucester, in consequence of the attack on Andrews, and arranged for a suitable number of men to put the work in hand at once. When he reached the stream which ran through he had been surprised to notice Tilly sitting on the bank fifty yards away, her back to him and her head sunk into her hands. It was her attitude which had chiefly impressed him then.

"Mind you, I only see her back," he said, "but I can tell you she looked desperate even from there, as if she was about at the end of everything. And I guess she must have been feeling that way, too, 'cos she got up all of a sudden, gave herself a kind of shake, and scrambled into the water just as she was. She might have been pulled in by wires, the way she went, and yet you could see she wasn't being pulled—she was doing it because she wanted to.

"I never reckoned that bit of water was deep, but it come up to her waist and higher, and I was still gawking when she stuck her blooming head and all under. I had a job getting her out, too: she fought like a wild cat till she went all limp in a moment, and as heavy as a ton of lead. Well, I lugged her back here, and that's all I know."

"And you don't know that," I declared, facing him squarely.

"Eh? What do you mean?"

He spoke slowly, a little suspiciously; but my smile dispelled any uncertainty about the way my words were to be taken.

"I mean, you saw her trip over a root and fall in, and you'd knock anybody's head off who suggested different."

His answer, I realized, was typical of the kind of man I had always subconsciously believed him to be: not crooked so much as pliable, never likely to miss a chance of benefiting himself for nothing.

"Would I, now?" he drawled.

"You would," I assured him, and rapidly pondered the size of the bribe it would be necessary to offer. I didn't stop to determine then why I should feel like bribing him at all, nor can I honestly say I thought the worse of him for his obvious invitation to be paid for any services I wanted.

We settled things without haggling at ten pounds, and he didn't even ask my motives, or seem to see anything unusual in the transaction. We also settled that unless he received contrary instructions he was to say nothing to anyone about the incident, having been asked not to do so by Tilly herself. A word with the cook confirmed my impression that there would be little to fear from her. She asserted with every evidence of good faith that she wouldn't make trouble for no one, let alone Miss Tilly, who always had a civil word, which was more than her father did, hoping I wouldn't repeat same, she was sure.

My talk with the girl merely bore out my instantaneous appreciation of the position: founded, if anyone is interested, on the manner in which Lina had asked for news of Tilly, and remembrance of the fact that according to Hinkson she was one of the chief suspects with regard to the theft of the ruby. Lina had been asking questions clumsily, and Tilly hadn't taken long to grasp that the stone was gone and herself under suspicion. It wouldn't need a genius to arrive at such a conclusion, especially as Lina wasn't cut out for the role of subtle questioner: yet I freely admit that on my part it was really only a wild guess which happened to be right.

"So you began to believe that everybody in the house but yourself must know the ruby was missing, and that most of them were quietly spying on you, and wondering what was the best way to make you return it. Possibly what you said to me the other day may have influenced you, too: and what with one thing and another you decided that heaven might be a nicer place.

"Of course, you're an ass, but I do sympathize, and I can tell you two or three things that may make you feel a little better. First, at present only Hinkson and Lina and I—apart from yourself now—know about the ruby; and a friend of mine at Scotland Yard, who's coming down tomorrow to try to find it before your father gets to hear. Second, I personally don't suspect you more than myself. Third, what really happened this afternoon was that you tripped over something and fell in the water, and Buck bravely jumped in and pulled you out, and you've asked him not to say anything in case people laugh at you for being club-footed, and he's agreed not to. That about covers things, I think."

She had listened to me in silence: which she now destroyed by bursting into tears. I did my best to cheer her up, and left her eventually sitting up and trying to smile, her nose red, her eyes bloodshot, her hair worse than anyhow, and her appalling green dressing-gown still clutched tightly round her. Before I went I obtained her most emphatic promise not to do any more silly things for at least a week, and I made sure she knew the story she must tell if anybody asked questions.

She said she was already feeling better, and might come down for dinner if she thought she could face it. I agreed that this would help to avoid undue notice, and ascertained that she possessed a second pair of glasses, so that the loss of the ones she had worn that afternoon need cause no comment. I didn't offer to stride off and fish in the stream, though: there are very definite limits to my chivalry. Incidentally, I don't propose to enlarge upon my reasons for being chivalrous at all. It's enough that I was sorry for the girl, and saw no necessity for standing by and watching her flounder if I could help without overmuch exertion. If it be thought that ten pounds was a lot of money to pay for the whim of assisting a totally unattractive female, I can only say that fortunately I could well afford it.

My original intention, formed as soon as I had verified my guess-work, had been to go for Lina hot and strong: but I soon realized that this was impossible, in view of the story that both Buck and Tilly were primed with. Regretfully, therefore, I exuded ignorance of the *contretemps* at dinner that night: but before the meal I wrote fully to Beale, telling him all that had happened since our telephone talk.

When the gong went at eight o'clock we all turned up in force that evening; and in form, too, if it comes to that. The conversation was more varied than usual, with Montague chaffing Adair and Hinkson on their self-imposed exile from the pleasures of life, and the Major rather brutally enquiring what a man could know about pleasure whose actions were dictated by a cheap inside. He was like that, capable at any moment of surprising insensibility to others' feelings. He surprised me also by telling everyone about Beale's forth

coming visit, and advising us jocularly to look in our cup-
boards and cover up our several skeletons. This went a little
near the bone with Lina and Hinkson, I fancied, and I could
see the girl tense herself against a possible enquiry about
the ruby. It didn't come, though, and I just heard her sigh
as Adair branched off on to the question of his progress with
the translation of the parchment clue.

"Still slow work," he said, with a wink at his secretary
which might have meant anything. "We hope we aren't
altogether wasting our time, though."

"And do you now retire to snooze again in private?"
asked Montague, when the meal was over.

"No fear," declared the Major heartily; and I knew then
that he must be up to schedule. "No fear, my boy: bridge
tonight. A good hard game—sixpence a hundred and no
revokes."

"Oh, I never do that," Montague said.

"No? What about Cambrai 1916, with Tenner and
Profitt?"

"Oh well, that was excusable. I was due for leave the
next day, and you could hardly expect me to concentrate."

"Cambrai?" echoed Lina girlishly. "Isn't that in France or
somewhere? But I thought you were disguised as a German
general all through the War, Nunky."

"Not all the time. Ladybird. I was dashing about all over
the place, and I don't think I was ever a general. A staff
captain once, because we needed the key to one of their
very special codes, and of course I know their lingo inside
out. Mostly they shoved me in front of a desk, though,
sometimes within earshot of the guns and sometimes not,
but always with the same job to do: pick other people's
brains. Bridge, Tony? And you might let me have a cigarette,
old man. And who'll be a fourth—Hinkson?"

"Well, not if you don't mind, sir. There's a concert on the
radio I'd rather like to hear, and the cook said she'd lend me
her battery set. And I must listen to the news, too. I haven't
seen a paper properly for ages, and one of these days you'll
ask me who's premier of France, and I'll be a week behind
the times."

"Ha! Just as you please—provided we can make up a four
without you. But of course we can. Tilly, fetch the cards,
and for God's sake cut out those damn silly psychic bids if
you play with me."

He scowled at her momentarily, and looked away to me.

"Go three of your opponent's best suit to show you've got a game in anything your partner's got ten of," he confided. "Something like that, anyway—never heard such rot."

In point of fact, however, as I think I mentioned earlier, bridge was one of the things at which his daughter got a chance to shine. Her afternoon's experience didn't seem to upset her game in the slightest and she carried me at least shoulder-high to victory in three consecutive rubbers.

Unfortunately, this achievement landed the poor girl in still more trouble. Her father was not a good loser, and though he had begun playing in what was for him a very good humour, he was soon frowning, and the frown had been steadily deepening into a ferocious glare.

"Devil take the cards!" he muttered sullenly. "Never saw anything like it. How much does that add up to, Roger?"

It was one of his idiosyncrasies never to keep the score himself, and yet invariably to insist on going through it afterwards item by item, querying everything he couldn't remember off-hand.

"Three thousand three hundred, eh? Let's have a look. Five hundred—now what in the world did they get that for? And eight hundred and forty as well—explain."

Montague did so; I did so; Tilly did so; and at last he subsided with bad grace.

"Well, one more," he said, though the time was after 11.00, and he had earlier declared his intention of being in bed by half past ten. "Yes, come on, one more rubber. Whose deal is it? And what are we supposed to be playing for?"

He addressed me, suddenly, as I shuffled the second pack while Tilly laboriously shared out our cards. Dealing was one thing she was no good at: it was painful to watch the precise way in which she slowly built up the four piles, never letting go a card until it was well in contact with the table or its fellows. Adair regarded her now with a baleful eye as he waited for me to answer.

"You suggested sixpence a hundred," I said.

"I did? Sixpence? Must have been mad. My God, girl, you make me dizzy. Anybody'd think the things would explode if you dropped 'em. And absolute rubbish," he added, picking up his hand and giving it a perfunctory glance. To his comments, she said nothing; but the colour came into her puffy cheeks a little.

I decided definitely during that last rubber that I did not like Adair. Perhaps it was rather late in the day to make up my mind, but I can't help that. To see a man holding bad cards in his own house tells you a lot about him: almost as much as to see him holding good ones in somebody else's. I think that at one period, to his credit, he made an effort to control his feelings, but before long they evidently grew too strong for him. Admittedly he was cursed with more twos and threes than anyone would care about who wasn't going *misere ouverte* at solo, but he needn't have flipped them down as if each was an invitation to a duel with poisoned épées.

"Sixpence a hundred!" he growled without warning in the middle of a hand I was playing, a tricky contract of four spades doubled. "And why the blazes you need a light on in the opposite corner of the room I don't know."

As he spoke he glanced savagely across at a finely wrought bronze lamp standing on an occasional table: an oil affair, as of necessity was all the illumination in Mauberley Grange.

"Waste of good wick and paraffin," he went on. "A paralytic blindfold imbecile could do what you're doing."

"Possibly," I said, a little nettled: "but one can do other things in the dark besides play winners."

"Such as?" he queried gruffly, though he knew my meaning: not that I was accusing him of any real desire to cheat, only I felt I didn't care much what I said.

"Revoke, of course," put in Montague quietly, with a faint smile. "And we're not allowed to—hence the more light the better."

Adair merely grunted at that, rose abruptly, stalked over to the offending lamp, extinguished it, and sat down again. I think I've said that he was a mean man over some things, and those often of the pettiest kind.

What I had feared might happen duly came to pass when the time arrived for settling up. I was deputed to receive payment from Montague, four pounds seven shillings, while the wretched Tilly got only the promise that her father would square with her 'the next fine day it snows'.

I looked at her, and just for a moment I seemed to see a veil lifted. Her glance at Adair was compounded of disappointment, contempt, and a something in addition that might have been naked hate.

But if he could be unpleasant, so could I too upon occasion. I repeat, I was possessed with an unaccountable mood of verbal recklessness.

"On second thoughts," I said, "I think I'll settle with you, Major. If you keep the money, then it'll help pay for my friend's keep. He eats a devil of a lot of marmalade, and they say oranges are getting quite dear."

Roger Montague responded admirably.

"Right," he murmured at once, and handed his debt to Tilly. Adair remained silent, but I rather gathered from his expression that he had just made the same kind of discovery about me as I had recently made about him.

Tilly gaped, reddened, looked wildly from one to other of us, and was reassured by Montague's smile and pat as he folded the money into her podgy hand. For a second he had the appearance of an ascetic bishop giving her a blessing.

For myself, I went straight to bed, breakfasted the next morning before anyone else was about, quitted the house by half past eight, and remained in Gloucester until Beale's train arrived.

And so I come to the end of my share in this tale. The rest will be related by Chief-Inspector Edward Beale, a sound man if ever there was one, and my very good friend. He is forty-three or thereabouts, nothing much to look at, five feet eleven inches in height, and has grey eyes. So far as I know he's never tried his hand at retailing his own exploits in print before, and it will be interesting to see what sort of a show he puts up.

SECOND PART

XI

As Tony has just said, it now devolves upon me to give the rest of the story, and I shall do it as straightforwardly as possible. The reason for the change of narrator may or may not become apparent, but the reader can be assured there is one, and that is as far as I feel inclined to go in the matter.

I have read my friend's side of the tale, and find nothing in it with which to quarrel seriously. The figures don't perhaps stand out with three-dimensional clarity, but it must be remembered that he was able to describe only what he himself saw and felt. He could not play the part of omniscient creator among the children of his own imagination, and it would not have been fair of him to build them up in the light of what he learnt later, subsequent to my arrival at Mauberley Grange.

There is just one thing I ought to point out before I get started. I don't promise that I shall always give all my thoughts about the persons or events which I describe. If, for example, the statement of a particular witness is open to doubt, I shall probably give my reasons for suspecting as much; but if it is an obvious lie, then I may leave the reader to note the fact for himself. As well, I am not to be taken as infallible in my views and deductions. Where possible, however, I will indicate at what points I became aware of having previously made a mistake, even if I don't wholly disclose its nature.

As I had told Tony on the telephone, I wasn't at all reluctant to take a short holiday, but I can't say I set out with any intention of working overtime on the problem of the missing ruby. After all, it is hardly possible for a police-officer to be very unofficial about police business, and the proper person to instigate enquiries into the whereabouts of lost property is the owner. I had little doubt, sitting in the train idly considering the matter as a change from reading or staring through the window at the bleak hurrying fields, that my ultimate advice could as well have been written on a postcard as conveyed by word of mouth. 'Tell Adair—the loss will fall on him, and every hour's delay probably lessens the chance of recovery, for which he wouldn't thank you.' However, it hardly seemed likely that my advice would be taken, judging from what Tony had told me of Lina and Hinkson. The girl would play for time, hoping that the full treasure might be discovered before her misfortune, and in the excitement the absence of one ruby be overlooked.

That is the worst of attempting to theorize without sufficient data: one is apt to be so ridiculously wrong.

I arrived at Gloucester punctually, and was glad to see Tony's tall form waiting for me at the barrier. On our way to Mauberley Grange, a run of about fifteen miles, he mentioned his near quarrel with Adair the evening before, and said he hoped it would make no difference to my reception. He also explained his transaction with Buck more fully than he had done in his letter

"I suppose you'll say I was mad to do a deal with him?" he hazarded, to finish.

"Well, unwise," I told him: "though I shouldn't think from the sound of the girl that anybody could misinterpret your motives. Still, it generally doesn't pay to give people like Buck a handle to pull. Mind you let me know the minute he sidles up and suggests lightening your wallet a bit more."

"Of course, if he does—but I don't think he will, somehow. He's had good value for his knowledge."

"Very good value: and he may think that if you'll pay ten pounds to hush up an attempted suicide by a girl you're not supposed to care a rap about one way or the other, there's always a chance you'll be daft enough to go up another ten."

"Well, I wouldn't."

"Why should you? In fact, why did you?"

He shrugged at that, but said nothing, and I didn't worry him. From my point of view, until I had met her, the only interest I should have in Tilly's actions would lie in the light they might throw on the disappearance of the ruby. People have been known to do away with themselves because they were really guilty of some misdeed for which they have come under suspicion; and more frequently, I should say, than because they were innocent.

We turned into the drive by the ill-conditioned lodge at five minutes to ten, not having dawdled. It was too dark for me to see the state of the gardens, as the waning moon was not up yet; nor to derive any clear idea of what the house looked like save that it was large and sprawling, its roof ragged against the sky. We drove round to the garage, once a barn, put away the car, and carried my two cases indoors through the massive porch. There was nobody about in the hall, and only a dim lamp burning sluggishly on a table, so we went straight up the wide paneled staircase and along various passages to Tony's room, where I was glad to see a bright fire burning. The second bed had been put up, and fitted with brand-new sheets and blankets, and I felt a momentary spasm of gratitude to life in general that I possessed one friend for whom it wasn't too much trouble to buy me the wherewithal to sleep in and on. I know a lot of people with quite as much money as he has who would expect you to make do with your overcoat and a couple of cushions in an odd corner, and pretend you enjoyed it. There seemed every prospect that I might have gone farther and fared worse.

"Pull up a chair and get warm," Tony suggested, after I had had a wash. "It really is perishing here at night. I'm sorry for Buck out in the woods—I wonder he doesn't invest my ten pounds in a fur coat."

"Perhaps he has done," I said. "It might hamper his movements a bit, though—I shouldn't think he'd be quite so useful in a scrap."

"Oh, I don't think Buck would fight with his hands," began my friend: but at that moment, eleven minutes past ten by my wrist watch, there came with complete suddenness a hesitating tap at the door.

"Now who the dickens is that?" he demanded. "Come in!"

Immediately the handle turned, and a tallish spare elderly man stood framed by the massive oak supports: a man in full evening dress, except that he was wearing a black waistcoat. His right eye was covered with a pale pink shade, which gave him a grotesque half-asleep effect, and he carried a lamp similar to the one I had seen earlier in the hall, though somewhat larger.

"Hullo, Andrews," said Tony. "You've swapped shades."

"Yes sir," agreed the butler in a quiet deferential voice. "The Major objected to the one you chose, sir, and found me this."

"Really? I'm disappointed. Anything you want?"

"Well, sir, the Major sent me to see if your friend had arrived. I must have just missed you downstairs."

He turned to me then, and spoke a trifle louder.

"Major Adair would very much like a word with you in the study in private, sir, if it's convenient."

"Certainly," I agreed, rising. "See you later, Tony."

I followed Andrews along the passage, and half-way down a flight of stairs I had not yet seen; and then on an obviously deserted landing he paused, facing round with a marked change of manner. He seemed to have become unaccountably nervous, and I noticed that the lamp in his hand was inclined to be unsteady.

"Sir," he whispered, with some urgency, "you *are* Chief-Inspector Bell of Scotland Yard?"

"Beale," I corrected.

"Oh, I beg your pardon. Well, sir, what I said just now wasn't strictly true. There's something wrong downstairs, and I don't quite know what to do, and I thought perhaps you could help, sir. Every evening at ten o'clock I take the Major in some sandwiches and hot milk, but tonight I can't make him hear. Would you come down?"

"Of course: but mayn't he be asleep?"

"Well sir, I think I would have made enough noise to rouse him. And I can't find Mr. Hinkson anywhere—that's his secretary; and Mr. Montague's gone to bed, I believe, and I thought it would be better not to worry the young ladies, and Mr. Purdon."

"Yes?" I prompted him. "Mr. Purdon?"

He looked uncomfortable.

"I understood there was some slight difference of opinion," he murmured. "That is, I don't fancy the Major would much care for him to see him like that."

"Now wait a bit," I said. "Like what?"

At this he appeared even more uncomfortable, and still further lowered his voice.

"Upon occasion Major Adair drinks a little more whisky than agrees with him," he told me solemnly.

"I see. All right, lead the way. Have you tried looking through the window, though?"

"Yes sir. I could distinguish that there is a light on but it isn't possible to see in."

It may sound as if we wasted a lot of time in talking but this wasn't so in fact. Both of us were speaking quickly, and I doubt if we delayed much more than a minute.

The floor of the Long Corridor downstairs was heavily carpeted, but the upper ones had not been: yet I couldn't honestly have said that in the later stages of journey to the study my companion made less noise with his feet than in the earlier ones. Tony had been right in telling me that Andrews moved with unusual softness of step.

Outside the door I stooped for a moment, listening by the keyhole after I had made sure that it could not be seen through. Behind me the butler hovered attentively, making sundry and would-be helpful gestures with the lamp, holding it near my face and almost singeing my hair.

I could hear nothing, and could not dislodge the key whose blunt end protruded slightly, blocking even the faintest gleam of light. Next I knocked, but without response, and a turn of the handle showed that short of breaking in there was no means of entering from where I stood. I was reluctant to use force until I had made sure there was nothing to be learnt through the windows, however, and said as much.

"Very well, sir," agreed the butler. "I'll take you round."

On the way he explained more precisely why I should not be able to see anything.

"The window's one of those old-fashioned ones, all in little squares and quite high up. It's in a kind of recess, a bay as you might call it, and inside there's a shutter across level with the side walls. The Major had it put there yesterday, in consequence of an attack on me on Wednesday night, and as likely as not it's padlocked. At least, it was when I took him in his supper at half past seven."

I realised at a glance as I stood shivering in the cold air that if I had to force my way into the room I should be far better advised to do so from indoors than out. The window of the study abutted on to a small patch of grass, at approximately either end of which stood tall trees: beeches, I thought, by the spread of their bare branches against the dimly star-specked sky, though it was impossible to see with any distinctness.

All I could be certain of was that the only means of looking into the room from without would be by getting a ladder. A narrow flat beam of weak light pierced the darkness at the very top of the window, and diffused itself upwards into a paling wedge. It began about fifteen feet from the ground, and I guessed that it might be caused by the shutter of which Andrews had spoken not fitting quite flush with the ceiling.

If I had felt justified in sparing the time I should probably have made shift to look in somehow, although the butler assured me there was no ladder about as far as he knew: but it seemed a case where delay might be dangerous. In answer to my explicit question Andrews admitted that Major Adair had never before been so drunk as not to rouse at repeated knockings.

"Begging your pardon, sir," he said, "but oughtn't we to get in somehow? I may be all wrong—perhaps it's a heart attack. . ."

He left his sentence not completely finished, and mentally I supplied three other possible endings: accident, suicide, or murder. You see, I am a policeman, and a detective more or less specializing in sudden decease at that, and I rather tend to think in terms of violence than of natural causes.

"Very well," I agreed, indoors again. "It'll be your job to pacify him if nothing's wrong."

And, after three more loud bangs on the door, I prepared to break it open.

As Tony has already mentioned, it was not a particularly massive one, though all in one piece, without panels, and I felt it give at my second charge. I also detected by the nature of its resistance that as well as being locked it was also bolted top and bottom. My third attack must have burst all three fastenings, because it went crashing open on its hinges, set to the right of the handle, and I floundered in ungracefully.

XII

Now, it is a well-known fact that the eye can take in an amazing amount in an inconceivably short time. Speaking literally, I can say that I perceived the state of affairs—as they were, of course, not including what might have caused them—at a glance. To elaborate: I saw on the left-hand side of the square room a polished mahogany knee-hole desk, upon the far end of which burnt a large brass lamp with an amber-coloured shade. In a chair between it and the fireplace, leaning back towards the latter, slumped the figure of an elderly man with white hair. His left arm hung down out of sight, his other had fallen on to his lap, and on the brown hair-cord carpet nearby lay an automatic pistol. He was plainly dead, shot through the mouth, and he was not a pretty sight with his glassy eyes wide open, seeming to glare at the ceiling, and blood all over his chin and waistcoat. His olive-green tweed suit was rather shabby, and even in the instant of regarding him I remembered what Tony had said about Adair's partiality for din-ner-jackets in the evening.

The fire still glowed redly, though it seemed not to have been attended to for some time. On the marble mantelpiece above stood a heavy crystal decanter about a quarter full of what looked like whisky, and near it an empty tumbler to match. In front of me, in a line at right angles to the normal closed position of the door, was a wooden partition formed of two flaps painted white, and these were fastened with a padlock. Thus the room had the effect of being entirely without a window, since the fourth wall, opposite the desk, was almost filled by a huge book-case. The partition ap-peared to have been newly made, and not particularly well, for though firmly enough fixed to the side walls it failed to fit along the top, as I had guessed.

I had the impression that the far corners of the room were full of books in toppling piles. To a height of perhaps five feet from the ground the walls were paneled in black oak, only the top row carved, and I also noticed that to the left of the hidden window hung a gilt-framed unglazed oil painting: a portrait of a man in sombre colours, head and shoulders only, about a third life-size. Otherwise that wall was bare.

All that, I say, I gathered into my mind, or wherever one's visual impressions are recorded, in an extremely brief space of time. If I had to give it a measured duration I should say that it could not have been longer than four seconds. In case anyone doubts that I could possibly have seen so much, I ought to say perhaps that I have trained myself to take in more than most people in an equal time. For a detective, quickness of observation is often as important as thoroughness. To explain why I made no move but a slight turn of the head for as long as four seconds, I think it will be enough to state that I knew without doubt that I was regarding a dead man, whom neither I nor anyone more qualified could have recalled to life by any imaginable swiftness of action; but put it down to the slowness of my reaction if you prefer.

If I have made that part clear, I can go on. At the end, as I judge, of the four seconds immediately following my entrance into the room there came abruptly from without, beyond the white partition, the cracking report of a firearm, succeeded a second later by another, and then a third. The noises seemed to me to come from considerably above ground level, and I instinctively moved forward a couple of paces, forgetful for a moment that I couldn't possibly see anything. Then in the ensuing silence, intense by contrast, I heard an exclamation behind me, and turned.

It was Andrews, two yards or so behind me to my right, and he both looked and sounded as though he were terrified. By his position, one hand still on the outer door-handle, he had been putting his head round to see if there were anyone in hiding in the corner, and I mentally congratulated him on his quickness of wit: but if he had been cool to begin with, he was cool no longer. He stood facing me for a moment, his left hand clapped to his ear and his pink eye-shade still incongruous, and then without a word of excuse he fled from the room, his coat tails flying.

Inappropriately I wanted to laugh, but instead I yelled to him to come back, and reluctantly he did so.

"You won't do an atom of good by rushing off outside," I said,—not that I really thought that his aim. "Let whoever it is go on shooting. Buck will deal with him—unless it is Buck. Stay indoors—it's safer."

I spoke at such length to calm the man: at closer view it seemed evident that this second shock had almost panicked him. First the sight of Adair's disfigured corpse, and then shooting outside within the space of five seconds: it wasn't really to be wondered at that he should feel like running away, especially as he didn't look at all the phlegmatic type.

Now, he might have made several different answers, if he had elected to say anything: assured me he wasn't scared, or made it plain in words that he was, or told me he was going to see that the doors into the house were locked, for instance. What he did in fact say surprised me, however. It was only one word, but it held enough meaning for a speech.

"Safe!" he echoed: and it was my first intimation that Adair's death could possibly be considered as other than suicide. Not even the sudden outside shooting had made me think of anything else.

On the face of things, of course, suicide was what it definitely seemed to be: but I have long ago given up the bad habit of being unduly impressed by surface appearances, and even without the butler's pregnant exclamation I should still have taken my next step.

I posted him at the door with orders to admit no one until he had consulted me, and then proceeded to examine the condition of Adair's study with great care.

I had better say at once that there was nothing in the remotest degree suspicious to be found, and for that reason I won't bother with a detailed account of everything I did. Naturally this included a certain scrutiny of the body and its clothes, unpleasant but necessary, the desk, the pistol, though without disturbing possible fingerprints, the decanter and tumbler on the mantelpiece, and a brief survey of the rest of the room, including the white-washed ceiling with its beams of magnificent ancient oak.

I satisfied myself upon the following explicit points.

Entry into the room, in the condition in which I found it, was impossible except by breaking in through window or door. The partition was genuinely padlocked, and would also have had to be somehow circumvented, and the door had been genuinely bolted top and bottom as well as locked. The latter fact alone might not have amounted to much as a barrier, since the lock was frail and old-fashioned, but the bolts were new, eight-inch solid brass ones, and my charges had torn the fastenings into which they slid bodily from the wooden side-beam.

Similarly, escape from the room was also impossible. I mean, nobody could have gone into the passage and from there have shot the bolts; or have departed by the window and yet left the padlock closed. I found the key to it in one of Adair's pockets, on a bunch with several others, and proved for myself that it was not of the self-locking variety.

(There is no trickery about the statements in the two preceding paragraphs, incidentally. Not with all the string, bent pins, lengths of special wire, or like novelists' contrivances for performing the impossible, could any mortal have sealed the room, from without, as it was indubitably sealed: the bolts were far too stiff.)

As far as I could tell, there were no secret passages or trap-doors. I am not much judge of that sort of thing, though, and referred the question mentally for expert opinion. I am aware of what Tony has said about this matter in the earlier part of his narrative, and am bound to remark that I think he and Roger Montague were being more facetious than profound. A detective tends to trust experts, if not by instinct then by necessity. He employs one to analyse the dirt under a suspect's finger-nails, and another to verify that a particular bullet has been fired from a particular gun; one to tell him from an examination of a corpse's intestines the probable dose of poison from which it died, and another to find from a fragment of spectacle lens the formula used by the oculist who ordered it from the maker. If he should set out in an attitude of disbelief, he would soon be reduced to despairing impotence. Personally, provided I'm confident that whoever advises me does or should know his job, and has no private axe to grind by misleading me, then I accept what he says until somebody else proves him wrong. The only experts of whom I am always a little sceptical are those of the medical profession, forensic branch, those who deal with handwriting, and those who make believe to foretell the habits and achievements of horses. The reason for my scepticism is that members of these learned trades and followings, more than of any others I know, are apt to contradict one another with sublime emphasis.

To return to Adair's study: entrance or exit by way of the chimney was impracticable: the fire precluded either.

Only one shot had been discharged from the pistol on the carpet, a .25 Webley automatic, and the ejected cartridge-case I soon discovered also on the floor. In my opinion the damage to the head was consistent with the idea that this one shot had killed Adair. ('Suicides' have been met with before now who displayed neat holes in their foreheads, but lay quietly clutching .45 Colts. Only an in-vestigating imbecile would be taken in by such a feeble ruse, however.)

Furthermore, everything pointed to the presumption that Adair had fired the pistol himself. It was lightly be-spattered with blood, and the end of the muzzle was still moist where it had been in contact—I felt sure—with saliva from his mouth. The backs, of his hands, especially the outsides of his thumbs, also bore marks of blood, and it was easy enough to reconstruct the probable shooting-position of the weapon: grasped with the hands together, and one or both thumbs on the trigger. If death had occurred in that way, then there was every expectation that the pistol would not remain in his grip. To support the idea was the disten-sion of the cheeks, and a mild splitting of the skin round the mouth, caused by the explosive gases generated by the shot.

I judged from the feel of the body that Adair had been dead not much more than an hour, if as long. From the lack of an exit wound in the back of the head it could be as-sumed that the bullet had lodged in his skull. This was perhaps a little surprising in view of the initial velocity of a missile from an automatic pistol—much greater than in the case of a revolver, being for a .25 Webley round about 1000 ft. per second: but there was nothing at all suspicious in the fact.

There were no recognizable hiding-places in the room: no cupboard or convenient curtain, no sofa to crawl under. The one spot where anyone could have decently concealed himself was behind the partition—though I repeat that he could not have done this and left it fastened on the other side. I easily determined that there was no one there, and I scarcely needed Andrews' assurance that there had been nobody behind the door. As well, I paid careful attention to the windows themselves. Only the bottom half was cur-tained, with a kind of net material, and I was soon rea-sonably satisfied that they had not been opened for some considerable time. The fastenings were stiff, and there were various other safe indications, such as the presence of many spiders' webs.

The condition of the desk was tidy, but not obtrusively so. As I have said, the lamp, stood on the corner farthest from the door, its mellow light leaving in shadow the top of the room, and illuminating at all brightly only its immediate surroundings. In the middle of the desk were two glass trays for pens and pencils, three bottles of coloured inks, and a rather tarnished brass bowl about eight inches in diameter, containing a miscellany of things such as rubber bands, clips, erasers, and rolls of adhesive paper. There were no fingerprints visible, because the top of the desk was covered with worn black leather, and there was plainly no last message.

It occurred to me then that there was no sign of the parchment Tony had told me about, nor of Adair's attempted solution. I found the former in one of the drawers, however—there were three small ones on either side of the knee-hole, all locked, but the key on the bunch: but still no solution. I glanced at the fireplace more closely, and now saw that there was a certain amount of paper ash on top of the coal, broken almost into powder. That might be the explanation, I thought, and rescued as much of it as I could.

I looked also at the tumbler, which was moist at the bottom: it seemed by the smell to have been used for whisky, but I was careful not to touch. Both tumbler and decanter, and their respective contents, would be removed for most thorough examination. The general impression from desk and grate and mantelpiece was that Adair had put his things in order before quitting their ownership, burned his attempts at solving the treasure clue, bolstered his courage with a last drink, and then shot himself.

I managed to go through the dead man's pockets without unduly disturbing him, though finding nothing else of interest but a small pocket diary which I retained for future scrutiny. First, though, I made a mental picture of his exact position, and desired Andrews to do the same: three eyes are better than two. I was confident that nobody could have tampered to any extent with the body after the shot was fired, though it would have been possible to drop the keys into his right-hand coat pocket, where I had found them. That in itself was not of course suspicious: nor the fact that the drawer containing the parchment would open six inches without coming in contact with his chair or leg. In normal circumstances, naturally, both body and room would have been exhaustively photographed before being touched at all, since life was so obviously extinct: but I felt justified in interfering on my own. For one thing, it would be some hours before a suitable camera and flash-light accessories could be procured, and for another, I had my own ideas about the way Adair's death should be investigated.

It was plain to me, even at so early a stage, that if I were right in assuming an absence of secret passages, then there would probably be no reasonable grounds within the room for suspecting foul play. Therefore, if in spite of that the man *had* been murdered, it must have been done with exceptional forethought and ingenuity—I speak in the mildest possible terms. That being agreed, it was equally clear to me that, in case there should be grounds for suspicion outside the room, my primary move must be to seem deceived by appearances. If I had to condense the whole business—vulgarly, art—of crime detection into a single axiom, it would be this: 'Never let a possible suspect know what you really think, but always let him believe that his expressed opinion, whether in words or deeds, as the faking of a suicide, is also yours'. In short, train yourself to be so convincingly insincere that if all else fails you can walk straight into a theatre and get your living by acting openly. I admit there have been good detectives who couldn't dissemble for nuts, but I think they would have been even better if they could have allied art to science.

In case I seem to have been cock-sure that because I was a Scotland-Yard man early on the scene I should automatically be given charge of the enquiry, let me say that I happened to know the chief-constable of the local force quite well. I was confident that when I had explained the position to him he would give me a free hand: that is, arrange with my superiors for me to work temporarily under his banner. If I had had to deal with a stranger, of course, I should have behaved very differently. Local talent is often reluctant to give up the chance of achieving local glory, and by butting in before I was asked I should have risked landing myself in fairly serious trouble. But I knew George Relf well. In truth, the fact had had its share in making me decide to come down to Mauberley Grange at all, and I wasn't afraid of his carpeting me.

XIII

I SPENT perhaps a quarter of an hour inspecting the body and the room, working quickly and quietly, and during the whole of that time I was on the alert for some sequel to the outside shooting which had so rapidly succeeded my entry. Andrews remained continuously at the door, except when I got him to impress Adair's position on his memory, but no one came near, and I was driven to the conclusion that the reports could not have been audible from Tony's room at any rate, and so possibly not from anyone else's. It would not have been like him to sit idly by and do nothing if he had heard.

I now sent Andrews to fetch him, and on his arrival made the position clear in a few words. He was most astonished and shocked, and assured me he hadn't noticed any shooting. He took a long look at the Major, still motionless in the chair before the desk, turned a little pale, and shrugged his broad shoulders.

"Now why in the name of sanity should the man go and do that?" he asked. "It doesn't make sense, Ted."

"We'll see," I told him. "Maybe he'd deciphered that parchment, and found it was a hoax."

"In which case he'd have torn the thing into shreds, not folded it neatly and put it away."

"Or perhaps he worked too hard and had a brainstorm—though possibly the same objection would hold. By the way, is that the pistol he lent you the other night? Don't touch, though."

Tony knelt down by the desk, and then looked up with a nod.

"I'm certain it is," he said. "I remember that chip on the handle."

"Thanks. Now, the immediate question is what to do. I don't fancy telephoning the nearest police-station, somehow—I'd rather go straight to headquarters."

I then explained my standing with George Relf, and asked if I could borrow the Bentley while he guarded the room for me in my absence.

"Of course," he said. "But what about all this extraneous shooting? And telling people? Tilly at least ought to know—he was her father."

"Yes, I suppose she ought to," I agreed. "And I'd very much like to be told where that secretary bird is before I go."

But even as I spoke that point was cleared up. Hinkson came rushing down the passage, his clothes somewhat disarranged. His stiff shirt-front was frayed and dirty, as if he had been scrambling through a moist hedge, his trousers were torn and stained, and his jacket and patent-leather shoes were similarly soiled.

"What's up?" he gasped—he had obviously been running hard. "Are you Beale? I've had a devil of a—hullo! Who's been busting doors?"

He paused and stared from one to other of us—Tony, myself, and Andrews a few feet away. For answer I silently bade him look inside the study, and his exclamation of surprised horror sounded entirely genuine.

"Good God!" he said in a low voice. "Murder!"

"Now why do you say that?" I asked immediately. "As a matter of fact, the business has every mark of suicide."

"Oh, rot!" he returned, a little shortly. "Adair would no more kill himself than grow wings and fly. And you haven't heard my story."

"No, that's true. How long will it take?"

"A few minutes. Let's go somewhere else, can we? I don't much like the look of—well, I'm so used to seeing him sitting at that desk alive that it gives me the creeps to think he can't move."

As he spoke he glanced for a second at the dead man, and then away in clear distaste. I solved this minor problem by sending Andrews for a screen, with which we hid Adair from view. I wasn't leaving the room unguarded for anybody.

All things considered, I thought Hinkson was taking his employer's decease in just about the way one might have expected, except perhaps that he showed a slight absence of surprise, and made hardly any display of regret. Still, that needn't be unnatural, I felt: from what I had heard of him, the dead man couldn't have been exactly easy to get on with. I studied the secretary closely as he talked, and came to the conclusion that Tony's description of him was pretty apt: on the tall side, slim, handsome, intelligent, and apparently well-bred. Admittedly he had been labelled as smart with regard to clothing, and now he was anything but that, with his soiled hands and shirt, and his black tie almost undone, but I looked for his story to explain the discrepancy: which it did. If I don't give it all in his actual words, it will be because I have so much else to relate that I can't afford the space. In fact, I'm rather afraid that parts of the ensuing narrative may be a trifle terse in style: but I promise not to gloss over anything of importance, and not to let it be duller than I can help.

Adair, said Hinkson, had not come into dinner that night. At half past seven, when the changing gong went, he had told his secretary to cut along, and return when the meal was over, and to inform Andrews on his way that he himself would have a couple of poached eggs on a tray. The reason for this departure from normality had been given without being asked for, and—from Hinkson's point of view—without its being necessary. There were only a few lines left to decipher, and Adair preferred to stick at the job till he was finished.

"Well," continued the young man, "I duly went back about ten to nine, and found him still sitting there working. 'Nearly done,' he said: 'only another half hour or so.' To be frank, though, I'm not sure that was strictly true. Although he tried not to let me see what papers and things he had in front of him, I don't believe they were the same as when I went out. I mean he seemed to have done with the parchment."

"Then what do you think he was doing?" I asked.

"I can't say. At a guess, either making a fair copy of the finished article, or puzzling out its meaning."

"The finished article? You mean the explicit directions for finding the treasure, not the whole document?"

"Yes, that's right—and I'll tell you another thing. The moment he got on to the actual directions, early yesterday afternoon, he set me to work on something else, with the result that I know nothing at all about the contents of the last two paragraphs—the ones that matter."

He paused, and looked at the desk, behind which was the screen hiding the body.

"I see you've been tidying up," he said. "Did he leave a message or anything?"

"No," I told him. "You see, I haven't been tidying up—the desk is precisely as I found it."

At that he stared in manifest surprise.

"As you found it?" he exclaimed. "But that isn't like him a bit. It's never been as neat as that—not five minutes after he moved in here, even. Then where's the parchment? Where's his translation?"

"I don't know," I said, not quite truthfully. "There hasn't been time to search thoroughly yet. What happened when you returned at ten minutes to nine?"

At first he was reluctant to go on till he had tried to find the missing papers, but I persuaded him that it was more important for him to tell me anything he knew.

He said that on his reappearance after dinner Adair had looked up from what he was doing, put his arms over his desk, and told Hinkson to sit down on the hard chair by the bookcase and listen carefully. He was to fetch his revolver, go outside, look round to see if he could note anything unusual near the house, and if not to climb somehow into one of the beech trees overlooking the study windows and the small lawn, and stay there watching. Adair's explanation for this peculiar command was as follows:

'I'm nearly finished, and I can't risk having anything go wrong now. Buck's somewhere out and about, but I'd rather he didn't know too much: that's why I want you right near the house, to make sure nobody can break in. If you see a soul stirring, make sure it's not one of the household, or that policeman fellow Tony's bringing down, and then blaze away. Out you go now: I'll tap on the window when you can come in.'

"So I went out," continued Hinkson, "taking his empty tray out to the kitchen on the way; and up, and sat there for a good hour, I suppose."

"Without even changing?" I asked.

"Yes. When I suggested it he got quite ratty, and told me not to bother. If the treasure turned up, he said, I could have a dinner-jacket for every day in the year, but there wasn't time to mess about now. I fetched my overcoat, of course, but found I simply couldn't climb in the thing, so I hung it on a branch, and it's probably there still. Anyway,. I hope so."

"Just one moment before you go on," I interrupted. "Did you get the impression that he seriously expected any trouble?"

Hinkson looked thoughtful.

"No, I can't say I did," he replied at last. "To tell you the truth, I thought it was just an excuse to get rid of me while he went through the translation of the parchment."

"I see. And similarly, you had no reason to suppose he contemplated killing himself? He didn't seem worried because he couldn't make sense of what he'd done, or anything like that?"

"No sir. If anything, he was rather pleased with things. Quite chirpy, in fact, for him."

"Thank you. And what happened after you'd climbed into the tree?"

"Well, nothing till about twenty minutes or so ago, when I suddenly heard a slight noise, and saw a man slinking round by the rhododendron bushes. I'd got used to the light, of course, what there was of it, but I naturally couldn't see what he looked like. All the same, I was certain by his manner and the way he moved that he was a stranger: he was thoroughly furtive, if you know what I mean. Well, I did what I'd been told to do, and fired three times."

I raised my eyebrows at that.

"A little rash, don't you think?" I suggested. "Supposing you'd killed the fellow? It would have been worth at least five years."

But Hinkson shook his head, with almost his first smile.

"Not really rash," he said. "You see, I didn't much take to the idea of maybe landing myself in clink, which was about what Adair was asking me to do if I did see anyone suspicious, so I took the precaution to load with blanks. Then even if I aimed at the wrong person—you, for instance—it wouldn't matter."

"I see. Have you the revolver on you now?" For answer he took it from his pocket and held it out: a six-chambered one of a familiar foreign make. Three of the chambers were still loaded with—as he said—blanks. I requested permission to take charge of it, and he offered no objection.

"And then," he went on "I got down from the tree as quick as I knew how, and was off after him—he'd scuttled away like lightning the moment I fired, of course. Unfortunately I didn't manage to catch him, though—ran into another tree and winded myself. But I met Buck—he's the bodyguard, so-called. He'd heard the shooting, and was coming to see what it was all about, and he's gone off to look for whoever it was I saw. I thought I'd better come back here myself, in case Adair was worried—and I find him dead. Do you wonder now that I immediately said it must be murder?"

"No, perhaps not," I agreed: "and yet I still don't see how it could have been, on the face of things."

I then hurriedly told him the circumstances of my discovery, and finished by popping a surprise question.

"If you were in one of the beech trees outside this room between about nine o'clock and twenty-five minutes ago" I said, "how was it you didn't see Andrews and myself walk round to find out if we could do anything from outside?"

He coloured abruptly, sat silent a moment, and then shrugged.

"I left that bit out," he muttered. "I went to sleep while I was up there, if you must know."

"When?" I asked. "For how long?"

"Can't say—how could I? It was after half past nine, because I remember looking at my watch then, and I woke again just before twenty past ten. What time did you walk round?"

"About quarter past," I said.

"Then there you are. I'm sorry I kept quiet about it, but I hate making a fool of myself unnecessarily."

"But otherwise your story's entirely reliable?" I queried, pointedly, and again he coloured.

"Entirely," he asserted, not without indignation.

"Very well—though you can hardly blame me for asking. One or two more questions: how long have you been employed by the Major?"

"Just over seven months—since the end of last May."

"How did you get the post?"

"In answer to an advertisement, in the *Times.*"

"That was after Warner had been dismissed?"

"Yes—naturally."

"Thanks. Now, when was that partition across the window put up?"

"Yesterday morning. It was an idea he had, but he didn't say much about it—just got Judd and myself to fix it. Judd's the chauffeur. I imagine the attack on Andrews on Wednesday night frightened him—have you heard about that?"

"Yes, thanks. Where did the wood come from?"

"Judd got it—made the thing himself, I believe. We put the bolts on the door, too, in case you're interested, but Adair produced those, and also the padlock."

"Did you have a key to it, or to the door?"

"No fear—had to knock like a little schoolboy every time I wanted to come in."

"And lastly, could you possibly have slept through the sound of the shot which killed Major Adair?"

That made him stare, and he didn't seem quite to know how to answer me.

"Oh no: at least—well, I shouldn't have thought so. But I certainly didn't hear it, all the same."

"Thanks—that was what I wanted to find out." I dismissed him then on a note of partial mystification, not letting him be too clear about what I thought of him as a witness. To be plain, I hadn't been impressed. I was left with a decided feeling that Hinkson's story wasn't by any means the whole truth, though which parts of it might need subsequent revision I couldn't of course be certain.

I now had to determine what I ought to do. It seemed hardly right to leave Mauberley Grange without informing the others of what had happened, yet the last thing I wanted was a clamouring crowd round the study. In the end I got Andrews to take me in search of Montague, Lina, and the dead man's daughter. They were all in their respective bedrooms, though only Tilly actually between the sheets. I introduced myself briefly, prepared her for a shock, and told her the news. She took it with a little gasp: yet showed thereafter less concern than I should have expected, even in view of her revelations to Tony.

"Oh dear!" she exclaimed, fumbling beneath the pillow for her spectacles. "Shot dead? That's terrible. Is there anything I can do?"

"No," I said. "At least, just one thing. Be kind enough to stay here in your own room, and I shall be much obliged. And you could also spend a few minutes in writing out a clear account of how you've passed the time since dinner, would you please?"

I made the same request of Lina and Montague. The former was curled up on her bed reading a not too respectable American magazine, and presented a vivid study against the black silk spread in a pink frock which showed off her admirable figure. She was an exceedingly pretty girl, as Tilly had struck me as being an exceedingly plain one, but she was spoilt by her petulant mouth and hard eyes. She received the news of her adopted father's death in a manner that bordered on the spectacular: first emitting the inevitable gasp, then sitting up with a jerky movement that revealed a startling amount of stocking, and finally clapping one dainty hand to her smooth white forehead in a gesture which reminded me of Andrews' in the study. Then she began to sob gently. Whether or not she was really upset I had no means of judging, but I fancied that she might be.

Montague, attired in a brocaded dressing-gown, was also reading, in an arm-chair by his fire. He rose at my entry, showing himself to be a frail man of medium height with a haggard face.

"Dead?" he echoed sharply. "How?"

"Shot," I said. "Apparently a case of suicide."

"Good heavens! But that's incredible. Is there anything to be done? I used to know a bit about first aid."

"It's too late for that," I told him. "He's been dead some little time—at least an hour, I should think."

I then asked formally for a written account of his movements that evening, and left him, returned downstairs with Andrews, who had been waiting for me on the landing, and told Tony I was going off to see the chief-constable.

"Right," he said: "and I stay on guard? I won't let anyone interfere. By the way, Buck's just come in—do you want to see him?"

"Yes, I may as well, I suppose."

I found the late Major's bodyguard in the dining-room pouring himself out a drink, took one look, and recognized the man. As it happened, I had been glancing at a photograph of him only the week before: one sent over from New York together with fingerprints and a detailed description, and the information that he was badly wanted across the water in connection with a case of kidnapping. Apparently he was English by birth but American by upbringing, and they called him 'The Cucumber' because of his disposition. They also supplied a string of names under which he might be passing, but neither Buck nor what Tony had reported as being his self-confessed real name, Kass, was among them.

I considered it a stroke of luck that I had the advantage of him: I knew who he was, but he didn't know I knew. For the present, understandably, I took good care to keep my

knowledge to myself. I merely told him who I was, established his Mauberley-Grange identity by a casual query,
learned that he had had his post for two months, but had
known Adair on and off for some years, and then asked for
any information he could give me. He held my eyes a
moment with his own bright dark ones, and then smiled,
disclosing his defective teeth. He also removed his slouch
hat to reveal an almost bald round head.

"Sure," he said, "but it ain't much. I guess you know
what my job's been here—prowling round at night and
keeping my ears and eyes skinned. Not that one man could
cover all the ground—what happened the other evening
proves that. Or at least, maybe it don't—I wouldn't care to
say. Anyways, tonight I was here and there and never saw
a thing worth a second peep, and then without a shadow of
warning I heard a couple of shots fired from round by the
house, and then a third. That was the direction all right, so
I pelted back and plumped right into Smart Feller. That is,
Hinkson, the guv'nor's secretary. Met him? Ah, then you'll
know what I mean. He told me he'd been stuck up a tree or
something cuckoo, and seen somebody creeping about,
and had a pot at him. We beat around for a bit, and then he
went in—he said he thought Adair might be worried. Time?
It was twenty after ten when I heard the shots, and I was
back here inside a minute forty seconds, which ain't bad
going from where I was. I know, 'cos I timed it. And now
they tell me the guv'nor's been croaked. How'd it happen?"

"If I knew," I said, "I'd be practically in bed. It looks
exactly like suicide. How would that strike you?"

"Suicide, eh? Well, maybe—why not?"

"Well, if it comes to that, why?"

"Ah, now you're asking things. But a man don't have to
have a reason for spilling his brains around. Maybe his
puzzle wouldn't work out—maybe he got the ear-ache and
couldn't stick it. My guess is as good as yours, if he didn't
leave a note."

"No, he didn't do that. Well, thanks for your help—would
you mind writing it for me?"

He looked a trifle suspicious at that. "What, shove what
I've just said down on paper?" he demanded.

"Precisely."

"All right, if you'll tell me why. Don't you believe me?"

"I don't know of any reason why I shouldn't," I told him. "It just happens to be part of my job to get written statements from possible witnesses whenever I can, that's all. There'll be an inquest, you know, and you'll probably be called."

"Oh, all right, if you say so."

But I could plainly see that he didn't like the idea, and I thought I knew why. He was afraid that his writing might fall into the hands of somebody who would recognize it. A similar fear, doubtless, explained why Tony had been unable to obtain his fingerprints. By now it was quarter to twelve. Before I got the car out I walked round to the back of the house and noted Hinkson's overcoat twenty feet up in one of the beech trees. I then tried to recall my memory of the reports, but couldn't settle definitely from what direction they had seemed to come, except that I had felt sure they were above ground. The location of an unexpected shot is a notoriously difficult task, especially without the use of one's eyes, and I decided that the three I had heard might well have been fired from where Hinkson said he had been.

I then made no further delay in setting out for George Relf's home a dozen miles away. The Bentley went well, as it always does, and at twenty minutes past midnight I was filling my pipe by the remains of the chief-constable's drawing-room fire. He listened in silence to all I had to say, asked a few questions, and readily agreed that he would at once apply to Scotland Yard for my services.

I won't bother with the details of the telephoning that followed: it's enough that when I returned to Mauberley Grange I was nominally working under Relf's supervision, but was actually in as full charge of the investigation as a mere chief-inspector can ever hope to be. Incidentally, I brought back with me another .25 Webley automatic, Relf's own, for subsequent tests, and while I was talking to the Yard I made arrangements for an enquiry to be set on foot concerning the present whereabouts of Warner, Adair's previous secretary. Relf, too, undertook to seek information about any strangers in the neighbourhood of late who might have behaved at all oddly. It was an obvious step to take, and I mention it only so that no one may think I overlooked it.

We agreed that the body had better be transferred to a more suitable place as soon as possible, and this was actually done, in passing, before breakfast the next morning. A car arrived at just short of three o'clock containing a police-surgeon, two constables, and one of Relf's brighter detective-sergeants. It was his job to make a thorough search of the study for fingerprints, and to take away with him the decanter and tumbler, and the pistol, for examination respectively by an analytical chemist and an expert gunsmith. He also took samples of all the different kinds of writing-paper we could find in the desk, and the ashes from the grate. His first task, though, was to photograph everything, and I had Andrews down to check the position of the body. The man was so sluggish, however, that I had to make him drink a glass of cold water before he was much use.

The doctor, a pleasant youngish man, couldn't tell me anything definite about the probable time of death, but said that as far as he could judge half past nine wouldn't be a wild guess. He promised to make a thorough autopsy without delay, and acquaint me with the result. He naturally departed with the body in the ambulance which had followed the car, and one of the constables with him, but the other stayed on in case of need. I told him to look after the study as soon as the sergeant had finished, and suggested that when it grew light enough he should go outside to look for footprints by the rhododendron bushes.

I then retired to bed with the written statements which Andrews had collected for me. None of them contained any fresh information, however.

XIV

AT this point I think I had better explain how I proposed to tackle the case. There was admittedly a preponderance of points in support of the suicide idea, but still several others to query, and my primary task would be to see if I couldn't dispose of some of them.

First, was a shot fired in the study, with door and partition closed, audible from any room in the house where others were supposed to have been between dinner-time and ten o'clock?

Second, ought the shots which I had heard fired from outside the study to have been noticed by anyone else in the house but Andrews and myself?

Third, how could Hinkson's account of his movements possibly be true? People without overcoats on don't easily doze off out of doors on a cold January night, and surely it was inconceivable that he could have missed the shot which killed Adair, discharged a mere twenty yards or so away? The reports from the beech tree had been audible from the study: then should not the one from the study have been equally audible in the beech tree? Or not? A matter for an acoustic expert, possibly, or of painstaking experiment.

Fourth, for what reasons could Adair be thought to have killed himself?

If because he had been unable to make sense of the parchment clue to Jasper Mauberley's treasure, would he have been likely to burn his attempts, put the parchment neatly away, tidy his desk, and finally shoot himself through the mouth?

If for some reason to do with the missing ruby, then Hinkson and Lina were both presumably lying.

If on account of some matter connected neither with ruby nor treasure, was it likely that he would have left no faintest hint as to its nature? (Suicides in general, in my experience, tend to prefer that the world shall know why they take their lives: in other words, they almost invariably seek to justify their actions. Either the reason is one about which their intimates can make a shrewd guess—love or money troubles, or illness, or a guilty conscience, or fear of exposure; or it is one which they think may be obscure, and accordingly they do their best to make it clear before they go. It is my own opinion that in excusing their contemplated deed to others the poor wretches, aware or unawares, are excusing it also and more urgently to themselves, working their minds desperately into a state in which they can feel able to ignore any other course but self-administered death.)

Fifth, what persons gained financially from Adair's death?

Sixth, what persons resident in Mauberley Grange could have had any other kind of motive for wishing him dead? (Three possible answers to this question at once suggested themselves. If Lina was the sort of girl Tony had described her as being, she might conceivably have stolen the ruby herself, and have later decided that she preferred Adair's death to the discovery of her theft. Again, almost anyone in the house except Tony could have come down to the Grange with the definite intention of waiting until Adair had cleared the path to the treasure, and then of removing him. In addition, Tilly at least was known to have private grounds for disliking her father—to put it mildly.)

Lastly, in the way of other general points for investigation, was the presence or absence of any suspicious fingerprints in the study; the presence of a known crook on the scene in the person of Buck; and the question of whether or not it was *possible* for Adair to have been murdered. (I mention this last so that no one shall think I was shirking it altogether. However, if you're to start working out the *modus operandi* of an apparent miracle, it helps to have some slight conviction that it may have happened through a human agency for understandable human motives.)

I won't pretend that I planned everything in detail as I have set it down here, but I had all the various items jotted down in my note-book, which I am never without, before getting up on Saturday January 22nd. I woke, surprisingly, at half past eight, and felt fresh enough to have risen at once, but didn't do so because I heard Tony say something about bringing up breakfast. I don't average more than one meal a year in bed unless I am ill, and jumped at the opportunity; not wholly through laziness, but also because I wanted a few extra minutes during which I could lie still and think out fully an idea which had occurred to me just before I fell asleep. It went, when I had elaborated it, more or less like this.

'There are ninety-nine chances out of a hundred that Adair's death was suicide: nevertheless, the odd one must be considered beforehand in case it turns up.

'If it should prove correct, and Adair *was* murdered, then the odds remain about the same that the motive was connected with the bulk of the treasure, and not with the ruby alone or some quite alien reason.

'If such should in fact be the hidden truth of things, then I must be ready to face the possibility that the murderer is even now in possession of the key to the treasure's whereabouts, and that he will seize the first favourable opportunity to get hold of it.

'Very well,' I told myself: 'if only as a ridiculously cautious precaution, I must take steps to see that no favourable opportunity can occur. The question is, how?'

I at once perceived several difficulties. Assuming a murder, and one or more murderers (with the probabilities favouring the plural, on the principle that what can't apparently be done at all will be done better by two people than by one), he or they might have struck from inside the house, or outside it, or from both together. For example: Adair had been killed by a member of the search-party working on his own, or by Warner, the dismissed secretary, or by Warner with the help of an inside accomplice. After consideration I inclined to the view that I need hardly trouble with the case where the whole thing had been done from without: there are limits to what I can swallow, even from myself. That being so, and the mechanical aspect of this probably imaginary crime being—I must suppose—wellnigh perfect, I should be hard put to it to produce a valid reason for turning down the obvious common-sense solution, suicide.

I mean, I should have next to no excuse for remaining at Mauberley Grange longer than twenty-four hours, or of seeing that any other official person did so except in respect of the man at whom Hinkson had fixed: yet to clear out, in the case of murder masquerading as non-disprovable self-slaughter, would be to provide just that opportunity which I was determined to withhold. I must therefore find authoritative grounds for staying on; but at the same time (for reasons I have already given) leave myself free to play the part of hoodwinked simpleton if necessary.

"Suicide, don't you think?" each one of the residents in the house might say to me; and to each I should prefer to reply "Undoubtedly—clearest case I ever saw." But to the further question: "Then what in the world are you still hanging about for?" I must have an answer, and a good one.

I found an answer, and because I could not know, lying in bed, that I should not need it, I proceeded to make the requisite plans for prolonging my visit, and so my investigation. While you're at sea you may still catch your whale, even if you have to float for a bit till you spot him, but the moment you're back on dry land you're sunk. And, as well as my own presence here, I wanted to assure that of everyone else. It would be no use for me to remain if a murderer were free to clear out till I had chucked my hand in, and then come back for the treasure when he pleased.

I took Tony into my confidence, with the result that at about ten o'clock he went off in the car to explain my suggestion to George Relf. The upshot, if I may anticipate, came about midday, due to violent telephoning on Relf's part. If I had been sure of being able to speak direct to Scotland Yard from Mauberley Grange in safety I should have done so, of course, and thus have avoided much subterfuge. For all I knew, though, thinking as usual in terms of the worst possible, a potential spy in the house-hold might have taken the precaution of bribing someone at the local exchange to keep watch on all outgoing calls.

Just short of noon, then, I myself received a call from Whitehall 1212, the gist of which I immediately spread abroad and half an hour later, in confirmation, there came up the weed-bound drive a telegraph boy, gingerly, a scared expression seemingly permanent on his freckled face. The moment he had delivered his missive he fled precipitately, and doubtless got straight into a strong dis-infectant bath at the first chance. The telegram read as follows:

Chief-Inspector Beale, Mauberley Grange, near Mauberley, Herefordshire. Confirmation of telephone call. Urgent. Woman you interviewed Thursday week Ealing since taken hospital smallpox. Consider house and grounds in quarantine. Local health officer already notified also chief-constable and railway authorities. Telephone Yard immediately this received. Vinney.

And that, I thought, would about settle things nicely. If it turned out that I was wasting my time I would get through to Relf for 'further instructions', and soon afterwards be informed that the diagnosis of small-pox had been incorrect, and the household was consequently free from restriction. To make things appear genuine, I pretended to be thor-oughly sick about the whole business, and apologized profusely to whomever I met.

One snag cropped up that I hadn't foreseen: everyone but Tony began to leave me as severely alone as if I had been visibly in the last stages of leprosy. That wouldn't do, of course, so I assured them all that if I were a small-pox carrier the whole house would be already infected, and those susceptible to the disease bound to catch it. Accordingly, there wasn't the slightest advantage to be gained from avoiding my presence. Luckily none of them knew if I was speaking the truth or not, and so by reluctant degrees they stopped edging away as soon as I came in sight.

The immediate effects of the quarantine order were these. The gates into the road were padlocked, and supplies ordered by telephone were to be pushed through or thrown over them, collected by Judd, and conveyed to the house. The constable who had arrived the previous night settled down to an indefinite stay, knowing nothing of the truth; but the two girls from the village, for whom there simply wasn't room, had been stopped at the bottom of the drive and told to go home. The health officer arrived soon after the telegram, fussed round everyone indiscriminately (he did know, of course), slapped us on the back with forced heartiness, and generally gave the impression that if he saw any of us alive again he would be very much surprised. His manner was most convincing, I may say, and he was of considerable help. In private he told me before he went that he thought my ruse had succeeded, though of course in a genuine case of isolation we should all have been vaccinated as a matter of routine. In this instance, he said, he had given everybody the choice, but there had been no acceptances.

On his jaunt Tony had made two other arrangements, the first connected with the local exchange. All our calls were to be dealt with by an approved operator, so that no question of interference could arise. As well, there were always to be two constables stationed in the woods, ostensibly to look for the man Hinkson had shot at; and one of them would carry any message I wanted to send Relf by hand, travelling by motor-cycle and bringing back an answer if required. It all sounds very childish, I dare say, but I could think of nothing better.

But why bother at all? Well, I have tried to explain, and if I am held not to have done so convincingly, I will bring forward two further points to justify my actions. First, in view of the attack on Andrews and Hinkson's mysterious slinking stranger I was, I thought, entitled to consider murder as a possibility until I could conclusively demonstrate that it was out of the question. Secondly, what I will call tradition was against the suicide solution.

It is a common but inaccurate supposition that no policeman ever reads a detective story without sneering. I have done so, and doubtless shall do, by the score, and have often found them not only interesting but potentially helpful. It's all very well to deride the amateur criminologist for getting his police routine wrong and working backwards from a made-to-measure solution: the one is only to be expected, since he is not a policeman, and the other is an essential condition for the production of his work. What is vastly more interesting is his approach to the most important problem in detection: that of how to tackle a problem; and there he is in the same boat with us—he has either to get on with the job or go out of business. Furthermore, as this problem is more often than not a matter of common sense rather than of special skill, and as common sense isn't confined to the ranks of officialdom, a policeman must surely have a very stubborn and unreceptive mind if he learns absolutely nothing from any dozen detective stories picked at random.

Now, detective fiction—in which I emphatically don't include mere thrillers of the *Bulldog-Drummond* type—has perhaps more and more rigorous conventions than any other kind, and one of them is that no death in a sealed room can possibly be suicide. It is always and inevitably murder, and the more certainly so the tighter the sealing. Nor does this rule lack sensible foundations: for it must be the ambition of every thoughtful murderer to achieve perfection, and what can be more perfect than to leave false evidence so convincing that there can apparently be no rational doubt about its truth? Yet the thing is potentially double-edged, because although what is desirable will sooner or later be attempted, and although we have all been taken in by conjuring tricks at one time or another, yet underneath in our minds we knew we were being taken in, and nothing in the world would ever have made us believe that the smiling damsel on the stage did in fact suffer bisection by the neighbouring circular saw.

The ambition of every true detective is to be un-deceivable, and the bogy of every detective is a criminal who shall prove cleverer than he. I too have this bogy ever at my elbow. Privately I call him Cain, and endow him with all the qualities I most fear to come up against in the course of my work: cunning, foresight, courage, and the ability to leave well alone. It is when things look most straightforward, when the solution of a homicidal puzzle stares me plainest in the face, that I begin to suspect Cain's hand behind the scenery, and although I have never found him yet I live in hope—or apprehension.

Now, Cain is the sort of man who would delight in a sealed-room murder, even if committed for no other reason than to confound me. I had therefore resolved, though exactly when I can't say, that I would test and re-test the suicide explanation of Adair's death, and not be content until I had eliminated every other possibility to my complete satisfaction.

XV

AND now, not before it is time, I must go back to the detective part of this story. Long before the house was officially placed in quarantine Relf's sergeant had departed, taking with him all movable objects from the study likely to bear fingerprints, and leaving me the definite information that none of the furniture or fitments showed anything suspicious. What prints he had found were either Adair's own or Hinkson's, which was only to be expected. The secretary raised no objection to having his impressions taken, and in fact the only person who did so was Buck. I could guess the reason for his attitude, but let him see that continued awkwardness on his part would make me not only impatient but vaguely curious about him, and he soon complied with my requests.

The next two points with which I must deal are my examination of the diary I took from Adair's pocket, and the meaning of my remark that my precautions against the cutting short of my stay were to prove unnecessary. I will deal with the diary first.

It was a small leather-covered book such as stationers sell by the thousand at Christmas, and it was scarcely written in at all. For New Year's Day was the entry 'Moved in'. January 2nd was blank, and the next thirteen days had each been given the one word 'Searching'. The 17th carried the information 'Found clue myself! Small box in door-post of panelled room containing parchment & ruby'. The 18th was inscribed 'Going to be hard', the 19th and 20th were blank, and the day of his death also except for a small tick in ink—the rest of the entries were in indelible pencil. That last might have meant anything, of course, but the most probable interpretation seemed as an indication of the fact that he had translated the parchment. Whether it could also be taken to include understanding of his translation was uncertain.

Otherwise the book looked disappointingly empty, and I might easily have missed the faint single line of lead-pencil lettering in the space allotted to August 7th. That I didn't is in no way creditable to me: I nipped through the pages rapidly and carelessly, and just happened to catch sight of it. This pencilled line was very short, consisting of three capital letters and a figure—MHE2.

For some minutes I stared at it uncomprehendingly as I lay smoking a pipe before getting up. Tony was by the fire putting on his shoes in preparation for going to see Relf, and he looked up as I spoke.

"What does MHE2 mean?" I asked.

He repeated the phrase, if that's the right word for it, and shook his head.

"A page in an atlas?" he hazarded. "Lots of them are divided into squares, with letters one way and figures the other. Heaven knows what MH stands for, though. Middle half, perhaps—whatever that would be; or mountains, high; or Middlesex Hospital, only that's W.1. Sorry to be so helpful."

"Don't apologize," I said: "I can't think of anything sensible either. But the queer part is that I've a feeling I do know what it means, if only I could remember. MHE2: now what can it be?"

He came across the room, examined the page of the diary, and then shrugged.

"The key to *Codex Codicum,* the Cipher of Ciphers," he suggested. "Guaranteed to interpret the riddle of the Sphinx, disclose the whereabouts of the lost ten tribes, and make the Book of Revelation as simple as a publisher's blurb."

"That's it," I said, almost without thinking. "It's a book—only, what book?"

He grinned at me as he pulled on his driving-gloves.

"Now don't you go taking everything I tell you for gospel," he warned. "But if it's a book you want, why stick at one? What about Macbeth, Hamlet, and Edward II?"

"Marlowe's History of Edward II," I murmured. "No, that isn't right: but something like, I fancy."

It wasn't till I had almost finished shaving that I remembered what MHE2 might mean; or at least, that I had used a similar abbreviation myself years back, when I was at school. The moment I was dressed I hurried down to the study and relieved the constable on duty. Before he went off for a wash and a meal he reported that he had been unable to find any clear footprints outside by the rhododendrons, though definite traces that someone had been moving about there.

When I was alone I turned to the big open bookcase opposite the fireplace. Its contents were extremely varied, ranging from most of the standard reference works to rubbishy novels and memoirs; from Brewer to the smallest of small beer, as one might say. By good luck I found what I was looking for within five minutes: volume II of Macaulay's *History of England,* tucked incongruously between *The Hound of the Baskervilles* and Kluber's *Kryptographik.*

My hopes soared high as I took it down, and were not chimerical. Inside, between pages 232 and 233, I discovered a folded sheet of lined paper from a cheap pad, thin enough not to attract attention when the book was closed except by accident. A glance showed me its importance, and I reproduce herewith what was written on it in ink in a small forceful hand resembling that of the entries in the pocket-diary.

And now for the gold & jewels that I have hid, & how a man must act who has hope in his covetous heart of obtaining them. Yet gold alone will not bring him to heaven, therefore do I call him covetous. Happy is the man that findeth wisdom, & the man that getteth understanding; for the merchandise of it is better than the merchandise of silver, & the gain thereof than fine gold. Wisdom is the principal thing; therefore get wisdom, & with all thy getting get understanding. Ponder the path of thy feet, & let all thy ways be established.

Let him take one pace to the north, & four to the east, & three to the south, & four to the west; & let him not be troubled that the measurements are inaccurate & the direction of no account. And for a sign, let him be polluted to the waist with blood.

'Well,' I said to myself, and wished it could have been to Tony, because I always find him an excellent person to talk to when I want to straighten my thoughts, 'some folk might think that was a first-rate reason for committing suicide: but would Adair have done so?'

I re-read the 'directions' carefully, copied them to the last detail—the right number of lines, the right words on each, and so on: and finally put the paper back between the same pages so that there was a space of exactly five millimetres between its top and the marbled edge of the book. That done, I returned the Macaulay to the case as nearly as possible in its former position. The shelf in question was five feet long, and while not crowded had perhaps no more than ten inches of spare room, made up of irregular gaps between the books. In view of certain considerations which had already occurred to me I proceeded to measure these gaps accurately with a pair of dividers from the bowl on the desk. I should thus have a fairly reliable means of telling by subsequent examination, first if the contents of the shelf had been tampered with, and second whether or not the paper had been removed from its hiding-place. My reason for these precautions will be found in point (9) of the list that follows. There seemed to be so many crowding my mind that the moment I was satisfied of having a check on the position of the Macaulay I sat down to clarify my thoughts in writing.

(1): Is the paper in MHE2 a true translation of the final paragraphs of Jasper Mauberley's shorthand parchment? If so

(2): Is it to all intents and purposes sheer balderdash, or has it a sensible meaning?

(3): Would Adair anyway have decided that it was drivel without spending a considerable time in studying it?

(4): If he had so decided, could he have been disap-
       pointed enough to be rendered temporarily insane by
       coroners' standards?
(5): Who put the paper in MHE2?
(6): If Adair, who else knows that he put it there?
(7): If Adair was murdered, was MHE2 overlooked by his
       murderer?
(8): Alternatively, is it possible that the person who killed
       him already possesses a copy of the translated di-
       rections, and because of their obscurity doesn't mind
       who sees them?
(9): Again, could a supposed murderer, having waited
       until Adair had completed his work on the parchment,
       and then having killed him (but HOW?), have found
       himself as much in the dark as ever, copied the di-
       rections, and then deliberately planted Adair's
       original manuscript in MHE2 in the hope that
       somebody brighter than he would find and make
       sense of it? If so, sooner or later he will look to see if
       the paper is still there, and I ought to detect that.
(10): Is the handwriting indubitably Adair's?
(11): Who in the house will admit to knowing that Adair
       ever had or kept a diary?

This also I re-read, and determined that for the present
nobody but Tony should hear of my discovery and speculations.

And now I will come to the second subject upon which I
must throw light: why I needn't have taken pains to ensure
an adequate length of stay for myself without making a
possible murderer suspicious. The reason is simple. Excluding Buck, who stuck to his opinions of the previous
evening, and Tony, who declared that he had none on the
matter (though he very soon developed strong ones),
everybody in the house alleged that he or she disbelieved
Adair would ever have killed himself, no matter what the
circumstances. Admittedly some of them later changed
their views, but by no means all, and in view of the circumstances in which I had found the body I considered this
general preference for the unlikely to be decidedly interesting.

I spent the rest of the morning, after my telephone call
from Scotland Yard gave me the excuse, in interviewing

people, beginning with Andrews. He was fifty-two years of age, he told me, and had been with Adair since June 1934. Also, he had served in the War, and had been slightly shell-shocked, and in an ashamed sort of way he proffered the fact as an excuse for his conduct the previous evening immediately following the revolver shots outside the study.

"I'm afraid I lost my head completely for the moment," he said, "but the truth is, nowadays I'm gun-shy, and many's the time the Major was on to me about it."

I told him not to worry too much about his misfortune, and asked for an account of his movements between the time dinner had finished and the time he attempted to deliver his tray of milk and sandwiches. He said that after clearing the table he and the cook—with Adair five years—had sat in the kitchen listening to her battery wireless set for the whole period. In answer to a further question he admitted that they had had the set on 'quite loud', because of the cook's partial deafness. Between 9.50 and 10.00 she had prepared the tray for him to take to the study, and this he had done as soon as the time-signal went. The cook, a Mrs. Farrar, corroborated his statements, and both assured me that they had heard no shot during the time in question. Mrs. Farrar, incidentally, had retired to bed at 10.05, and didn't learn till the next morning that anything was amiss. Both of them gave it as their considered opinion that their late employer would emphatically never have done away with himself; but I naturally didn't place too much reliance on what they thought about the matter. I finished by asking Andrews a few questions about the attack made on him on the Wednesday night, but learnt nothing fresh.

Next I sought out Tilly, as the dead man's next of kin. She was dressed, a little to my surprise, in a black frock: an ill-fitting garment that might have been borrowed from a parlour-maid of entirely different size and shape. She looked as if she had not slept decently for some time, to judge from the unhealthy lines visible beneath the lower rims of her spectacles, and she kept folding and unfolding her hands nervously as she perched on the edge of an arm-chair in the communal sitting-room. Otherwise her manner was very solemn and intense, and I could well imagine that nobody would ever choose her for a cheerful companion.

"Did you ever hear your father threaten to take his life?" I asked her after a bit.

"Oh no, Inspector, definitely not. He—he was not at all that kind of person."

I remembered then that the subject of suicide would hardly appeal to the poor girl, but did not immediately change it.

"Do you mean then that you don't believe he did die that way?"

"Yes, that's what I really think,"—with an owlish nod.

"But how else could things have happened?"

"I don't know, but I shall never believe he would have shot himself, never."

"And what makes you so sure?" I asked, but could get no very satisfactory reply. It seemed that she derived her opinion solely from experience of her father's character.

"I can't give you any convincing reasons—I'm just sure. You know how one feels about people after living with them for a long time: one realizes there are some things they would be simply incapable of doing. Oh,"—as I was about to speak, "I remember you said the room was locked, and nobody could have got in or out, but I can't help that. There must have been a way—or else it was an accident."

"I hardly think so," I murmured. "Had your father any worries apart from the treasure?"

"No, I think not. That is, not unless he had got to hear about the ruby's being missing."

"But you hadn't mentioned it to him?"

"Me? Of course not."

"When did you last see him alive?"

"At luncheon yesterday. He didn't come in to dinner, and we don't have a formal tea, and I shouldn't have dreamt of going along to the study unless he sent for me."

"And when were you in there last?"

"Oh, that's hard to say. A week ago, at least."

"I see. Two more questions, Miss Adair. First, do you know if he kept a diary?"

She shook her head.

"He had one of those stupid little things with no room in them—Tina gave it to him at Christmas; but I don't think he would have written anything in it. You see, I—I keep one myself, and he always used to tell me I was a fool to do so."

"For any particular reason?" I asked: and her answer made me think that Tilly was at last coming out of her shell now that her father was dead. She spoke bitterly—the more so to my ears because her tone was so very matter-of-fact.

"No, I don't think he had any special reason. Just because it was I who kept it, I dare say."

"And finally, do you know if he had made a will, and if so, where it is?"

She coloured considerably, and looked embarrassed.

"I only know he *said* he would make one, about a year ago," she told me.

"With what object?" I asked, and let her take the question which way she pleased. She appeared to think I wanted to learn why her father should have told her his plans, and her answer was disarmingly open.

"Well, I don't see why you shouldn't know, Inspector. My mother died ten years ago, and left what little money she had of her own to me, not to Daddy. They didn't get on at all well together, and had been separated since 1920. It wasn't a great deal, and not very cleverly invested, and for the last three years I've had hardly anything from the interest—fifteen pounds in one year was the most. In spite of that, Daddy was always very sarcastic about my private income, as he called it, and would never make me a proper allowance as he did Lina. He adopted her in 1935, and on my twenty-first birthday told me he was going to draw up his will, to make sure his money would go to the right people when he died. As I was already provided for, he said, he would be leaving her the bulk of his estate. He—he was a bit drunk at the time, and we'd been quarrelling most of the evening, because he said the champagne was cheap and nasty, and I explained that I couldn't afford anything better. So you see that I don't really know if he meant what he said then or not, but I think he may have done. He never mentioned the subject again, though."

There were two obvious points to query.

"Was Lina in the room at the time?" I asked.

"No."

"And you really mean to say that you were expected to provide champagne for your own twenty-first birthday party?"

"Oh yes," she answered: as though to do so were by no means unusual. If she had told the truth, and, there seemed no reason to doubt it from her manner, Adair must have been at times a most unpleasant person, I thought. I let her go soon afterwards, having secured the name of her father's solicitors, which she happened to know, and the date of her coming of age, October 16th 1936.

Next I interviewed Lina, whom I found in her room staring aimlessly out of the window at the fine rain which had just begun to fall. She too was arrayed in black, but her frock was a good deal smarter than Tilly's, and I felt sure that it had been originally bought because it suited her fair type of beauty, not for purposes of mourning.

My brief glance at her the evening before had done little to make me inclined to like her, and quarter of an hour's chat now did no more. She struck me as being all that Tony had said: vain, sophisticated in the worst sense of the word, a trifle common, and physically a goddess careless of her eyes and mouth.

I tackled her from a different angle.

"I'm sorry to be nosy, or badly informed, Miss Hipple," I said, "but would you very much mind telling me just who you are?"

"No—why should I?" she countered, staring at me rather sullenly, as if she preferred her own thoughts to my company. "You've got the name right, anyway, and I was Major Adair's adopted daughter. Or maybe I still am—I'm darned if I know where I stand."

"Then I take it that your own parents are dead?"

"Well, yes and no. My father is, but nobody knows what's happened to my mother."

She paused, but after a few moments of silence seemed to feel, correctly, that I was waiting for her to continue.

"My father was a friend of Uncle's—the Major's," she explained listlessly. "They lost touch after the War, though, and he died in 1933, and it wasn't till 1935 that Uncle heard about it, and hunted me up to see if there was anything he could do. I was living with an aunt in Bromley then because my mother had hopped it a year before—went off with some second-rate actor, and that was the end of her. As far as I was concerned, I mean. So the Major adopted me, and there you are."

"I see—thank you. Now, I'll just ask you three or four questions about last night, if I may. According to your statement you were here in your own room from after dinner until I told you of Major Adair's death?"

"Yes, that's right: from about five or ten to nine."

"Were you alone all that time?"

"Yes—of course."

"And you heard no shot?"

"None at all—not even the ones Dick fired."

"Dick?"

"Mr. Hinkson."

"Can you think of any reason at all why the Major should have committed suicide?"

"No. He couldn't have."

"Why not?"

"Well, he wouldn't have."

"Not in any circumstances? In your opinion, I mean."

"No, I'd say not."

"What makes you so sure?"

At this she shrugged, and lit a cigarette before answering. All this time, in passing, she had not thought to offer me a seat, and accordingly I remained standing: Lina did not appeal to me as a young lady with whom to be on any but formal terms.

"Oh, one thing and another," she told me vaguely, after expelling a cloud of smoke. "The way he used to go on about people who cut their throats or park their heads on railway lines. He said they were all cowards, and he couldn't stand cowards, and while there was life there was hope, and all the rest of it."

"Thank you. Now, when did you last see him alive?"

I put the question in that form purposely, and she quickly pulled me up.

"Now, none of that wife-beating business," she said, though not bad-temperedly; rather with a kind of smile that might have been due to nervousness. I wasn't sure—it seemed unlikely to be one of her failings.

"I haven't seen him dead," she added, "so the last time I saw him he must have been alive, mustn't he? It was at lunch yesterday, or just after in the hall. I asked how he was getting on with his old clue, and he said he might be through with it any day now."

I nodded.

"And one last thing," I remarked, preparing to go. "Have you any reason to suppose that up to the time of his death, say ten o'clock last night at latest, he'd heard anything about the loss of the ruby?"

The question clearly surprised her, and she seemed to need a moment to collect her wits.

"Why no, of course not. Why, that was what you were coming down for, to find it before he did get to hear."

"Yes, yes," I agreed, a little impatiently, "but there was a long gap between the time my visit was fixed up and last night: quite long enough for him to have asked to see the stone suddenly."

"Well, he didn't, and nobody told him because only Dick and Tony and I knew. At least—I'm not sure about that. Maybe somebody else did know."

She spoke her last words slowly, with a frown, more as if she were thinking aloud than talking to me.

"Somebody else?" I queried. "In particular?"

She frowned a moment more at that, and then laughed abruptly.

"Oh, you work that one out for yourself," she said; and I had very little doubt that she had been referring to Tilly.

XVI

THIRD on my visiting list came Montague, in bed again today. He admitted that he was not feeling especially unwell, however, in answer to my direct question on the point.

"But what is there to get up for?" he asked. "I can't do any good downstairs, and anyway I want leisure to think."

"Think?" I repeated, hoping that he would be communicative.

"Yes—about my future, you know."

As he spoke he smiled faintly, his lined face looking for a moment as if he might be no more than sixty after all. But his next words disclosed his real age, or what he chose to tell me was his age.

"I don't doubt you'll think it's silly for a man of sixty-nine to talk about his future as if he were an infant," he said, "only it happens to matter to me. I don't know if Tony told you anything about our little talk the other night—not that there's any reason why he shouldn't have done. The fact is that my bank balance is like my stomach, in a very poor state of health: and now the one possible doctor's gone and died. Yes,"—in answer to my expression, I suppose, "I mean Adair, Inspector, and you must think what you like of me. I'm not going to pretend to be overwhelmed with grief, or any of that nonsense, because I'm not, but I shall miss him, all the same. Adair and I were good enough friends once, but latterly we hadn't very much in common. Then why am I here? As I told Tony, purely for reasons of pocket, not sentiment. I did Francis a good turn once, and it was up to him to do one for me."

He stopped abruptly, leaving me to direct the conversation. But I could be as apparently callous as he, and there seemed no harm in meeting him on his own ground.

"You're not suggesting that the Major committed suicide just to get out of helping you, I suppose?" I said, and he smiled again, more broadly.

"Good lord no!" he assured me. "Francis may have been mean, but not so mean as all that."

"No, probably not. What's your opinion of his death?"

"Now, my good man, be reasonable. I only know what I've been told, and when I've got no straw I'm a proper Israelite. Apparently you found him in the study with a bullet in his head and a gun practically in his hand, and the room's alleged to have been locked and bolted and all the rest of it. One would assume suicide—and yet I'm not quite happy about it."

"No?"

"No. In fact, I don't like the idea at all. You see, I knew Adair pretty well, taking things all round, and I can tell you one thing about him that may have escaped other people's notice: he was a coward."

"Really?" I murmured. "Physically or mentally?"

"Oh, come now! What is a physical coward, pray? If you run away from barking dogs or dentists or cricket balls, you do it because you funk the *idea* of being hurt by them. It's your mind that's afraid, not your body."

"Obviously," I agreed. "Yet there are hundreds of people who fear nothing except the idea of being physically hurt. I call them physical cowards to distinguish them from people who're only afraid, for example, of what others say or think about them."

(A childish digression, I grant, but none of my choosing.)

"Oh, all right—then Adair would definitely come in the first lot. Yes, I see your point: suicide's traditionally the coward's way out, but how can it ever be the physical coward's way? Not that a bullet through your head hurts much, I should say: but then, you'd always be terrified you mightn't aim straight."

"Very well," I said. "In your opinion he didn't shoot himself. Then kindly tell me who did—or better, tell me how anybody could have done."

"In fact, do your job for you?" he suggested amusedly.

"Exactly—or give me full directions. How can you kill a man in a sealed room, and yet not be there when the room's broken into? I can assure you it would have been impossible for anyone to have left the study, from outside, in the condition in which I found it."

"And equally, of course, that Adair was the only person in it?"

"Of course. If you know the room at all, you won't need reminding that there isn't a single possible hiding-place that would stand a moment's scrutiny, and I can promise you I looked."

"Well, we won't quarrel about your use of the word 'promise'," he said. "A sealed room, eh? I seem to have heard about such things: synonymous with murder, aren't they? But is this particular problem so very hard? Or don't they make revolvers with a range of more than two feet any longer? Look—this is how I should have done it, or how I imagine it could have been done.

"Adair is sitting quietly at the desk in his study when somebody—myself, for instance—knocks, is admitted, and stands chatting for a few moments. Then, choosing my opportunity, I slip a dose of some quick-working but relatively harmless sedative into his whisky. Next I ask for a drink myself—he always keeps a spare glass in the study; wait till he's finished his, change tumblers, and hang about till it's almost time for the stuff I've given him to work. Then I carelessly drop the glass I'm holding, really his doped one, pick up the pieces, apologize, and clear off. He naturally—in view of the fact that he's who he is—locks and bolts the door after me; but he's feeling muzzy, and immediately sits down again. He then sees something I've left behind me, perhaps on the floor near his chair when he wasn't noticing: a revolver, I mean, with one chamber recently discharged. He picks it up, effectually fingerprinting it, gets still more muzzy, and soon passes out. Meanwhile I've slipped round to the window, and when I think it's safe I sneak in, creep up to the partition, and push the thing inwards till there's a crack big enough to peep through. I see him sitting unconscious in his chair, get out another revolver identically like the one in his hand, or on the floor if he's dropped it again, take careful aim, and polish him off. Then I retire through the window, and wait for the chorus to say 'Suicide!' "

"Which it hasn't done, strangely enough," I remarked.

"No? Well, that's only a trifle. Pick out the other flaws."

I smiled: though actually I was thinking hard, wondering if Montague had any purpose to serve in making out that he was unaware at what range Adair had been shot. It was difficult to get any indication from his face: even to my not inexperienced eye he looked honest enough.

"I wish I couldn't pick them out," I told him: "but it's too easy. In the first place, Adair was facing the book-case opposite the fireplace, at an oblique angle to the partition: yet he was found shot through the mouth. If I'm any judge, the muzzle of the pistol—not revolver—was actually in it at the time. Moreover, there's no such thing as a gun that's identically the same as another gun. Nowadays it's possible to prove beyond doubt whether a particular bullet came from a particular firearm or not, provided the bullet's in fair condition and the firearm hasn't been damaged. The science of ballistics, they call it."

"Oh, that!" he said. "I thought it was just storybook stuff."

"No, it happens in real life as well. The instrument used is a hastoscope."

"Then I've learnt something new, even at sixty-nine. And it seems that I haven't been as clever as I thought—though I didn't know about the shooting through the mouth."

"And who told you about the glass—Andrews?"

"Glass? Oh, for the drink. No, I took that for granted. It would have been an event to visit Adair after dinner, and not find whisky somewhere near. When he was alone, that is: unlike most people, he preferred his liquor in solitude."

I went away wondering two things: if Montague had had any reason for giving me so elaborate and unsatisfactory a theory of murder, and if he and Adair could ever have been very great friends. If so, the survivor of that friendship, even though it had cooled down the years, was hardly behaving in the usual *de mortuis* way.

I can't honestly say I cared for Montague: but then, I had yet to meet anyone in the house bar Tony who struck me as possessing any particular merit. Hinkson was inclined to be shifty, I thought, Tilly was dull, Lina an artificial minx, Buck an American crook, Andrews a soft-footed nobody, and there was only Judd to meet. You will please understand that these were merely my impressions after being in the house for approximately fifteen hours.

I was half-way down the stairs in search of Buck when I remembered another question I had meant to put, so returned.

"Who is Lina, Mr. Montague?" I asked: but he shook his head.

"Glamour *a l'anglais,"* he said: "that's the best I can do for you. I met her once before I came here, but I don't know anything about her previous history."

"I see—thanks. I had an idea that her father was a great friend of Adair, and as you were too I thought you might have met him, perhaps."

"Never—nor heard Francis speak of him. Nor Lina either, if it comes to that; but then, she isn't very talkative about herself—or only in the present tense. What makes you ask?"

"My insatiable curiosity," I said. "It's almost my master. And while we're on the subject, tell me something else: was there a waste-paper basket in the study the last time you went in there?"

"Sorry—I haven't the slightest idea," he answered. "I don't remember seeing one."

"No? That's queer. How long ago were you in there?"

"Oh, some days—Tuesday, I think it was."

As I turned to go the luncheon gong sounded, gently, as if out of respect for the dead owner of the house. We ate almost in silence, since it seemed that collectively Lina and Tilly and Hinkson were still not quite prepared to accept my supposedly germ-laden presence with any enthusiasm.

Afterwards I sought out Buck, who took his meals in the kitchen. The rotund asthmatic Mrs. Farrar informed me that he had just gone out, but wrongly: ten minutes later I caught up with him in what seemed to me strange circumstances, considering what I had been told and had noticed for myself. He was talking amiably in the lounge to the dead man's secretary, the two of them apparently the best of friends. I thought, though, that their mutual amiability was just a little forced: not for my benefit, but for one another's. Both smiled at my approach, and for a moment I feared I might be invited to join in a small circle of benevolent boredom; but Hinkson, with less than obvious tact, announced his intention of writing letters.

"I'm out of a job, you know, Inspector," he said, "and I'll have to do something about it. Of course, I haven't forgotten we're all stuck here for a fortnight, or however long it is you have to wait for the spots to come out, but if I don't go to hospital I must land some sort of post pretty soon. And how do we go on about letter-writing, by the way?"

For a moment I thought he might be trying to catch me out, but if so I was ready for him.

"I fixed that up with the doctor chap who came this morning," I said. "He's arranging for them to be called for every evening, and disinfected before they go out. I'm afraid it'll mean leaving the envelopes unsealed, but he promised there wouldn't be any snooping. All the same, it mayn't strike everybody as a good idea, so if you've any alternative suggestions to offer, I'll see what can be done."

"Oh, I shan't mind," declared Hinkson with a smile." Thanks for telling me, though."

"Not a bit—I ought to have done it earlier. Perhaps you wouldn't mind passing the news on, if you've a few minutes to spare?"

"Certainly. And while I remember, would you have any objection to going through the Major's desk with me some time? I can't help feeling it's odd he shouldn't have left his translation about, and I'd like to look. Luckily I've got my own notes of the first part, if his don't turn up, and I was thinking that perhaps during the time I've got to be here I might have a go at the last two paragraphs of the parchment. If there is a treasure we may as well find it, don't you think?"

"Well, if it was any of my business I'd say yes," I told him. "There's no possible objection to going through the desk, and I'd like your help, but until we know who inherits his estate I can't advise you where to go for permission to continue looking for other people's property."

"Hang!" he exclaimed. "I hadn't thought of that. But was it his—or would it have been his, rather? Isn't there some law about things like that?"

"Yes: but the parchment was certainly his, and without that you won't be able to do very much, I imagine. You don't happen to know if he left a will, I suppose?"

"No—I never heard him say anything about one. When you want me I'll be in my room, I expect."

The most interesting point arising from that conversation, of course, was the secretary's implied knowledge of the fact that Adair had not destroyed the clue as well as his translation of it, and the question was, how did he know that? I had certainly not told him myself. For a moment I thought I might really have spotted something important, but a little reflection supplied a possible answer. Asking Buck to wait a few seconds longer—he had withdrawn to the window and was glancing at the paper—I went along to the kitchen in search of Andrews. In answer to my query he admitted readily that he had heard me mention the parchment's survival to Tony the night before, and in the course of a casual interchange of remarks with Hinkson that morning had passed on the information.

'So there goes one swan that was really a goose,' I thought, as I returned to the lounge.

"A nice chap, Hinkson," I remarked to Buck. "Or not?"

He grinned widely at that, and for the second time I couldn't help wishing that he'd see a dentist, or at least buy a tooth-brush. Perhaps the matter would be taken out of his hands, though, I remembered. It would plainly be contrary to my duty to leave Mauberley Grange without arresting him for his alleged transatlantic exploits, but for the time being I had an idea that he might be more useful to me at large and unsuspecting.

"I dunno," he answered. "I usen't to like him better'n a packet of rat poison, but I guess I may have been all wrong. You know, I always got a kind of feeling he was all out for number one, but I'm durned if I see how he can have gained a cent by the guv'nor's death. I reckon maybe I was a bit hasty. Maybe it was his manner got me rubbed up rough, or his creases and fancy ties, or something. Still, you ain't here to talk about him now, are you?"

"No, not really," I admitted. "What I want from you at the moment is an opinion. Could anyone have known what you'd do between 9.00 and 10.00 last night? If so, who?"

Buck looked at me curiously, and pulled at his scanty hair.

"You put a hell of a lot into a few words, don't you?" he remarked, almost casually. "Anybody, how—who would that mean?"

"A member of the household. I ought to have said 'known beforehand'—I'm not counting people who crept round behind you in the woods, if there were any."

"There weren't," he averred confidently. "Nobody don't creep behind me without I know it. Well, I think the answer's no. If you were to stretch it a bit, and say between 7.00 and 2.00 or 3.00 in the morning, though, I might say yes. They could bet I'd have been everywhere in the grounds: but I wouldn't know myself just where I'd be for any special hour."

"Thanks—that was what I wanted. Can you put any limits to it? Could anyone have relied on your being, say, at least a hundred yards from the house?"

"Oh yes—two or three hundred, maybe. But near enough to hear shooting, if that's what's on your mind. I heard Sonny Boy do his bit."

"Yes, so you said. Yet you apparently didn't hear the shot that killed Adair."

"Well, that was indoors, wasn't it?"

"One assumes so. I wonder if you'd show me just where you were when Hinkson fired?"

"Why, sure, if it makes you happy."

Before we started out I warned the others, especially the gun-shy Andrews, to take no notice of any reports they might hear, and made certain arrangements with Tony about Hinkson's revolver and the pistol I had brought back with me from Relf's. At precisely 2.30 he was to fire three blank shots from the first, standing on the little lawn outside the study window. Then, five minutes later, he was to fire again from inside the room, this time using the automatic. I showed him how to work both weapons, and because Relf had only had real ammunition we rigged up a suitable target for his second display—an enormous log a foot thick, upon which he solemnly and optimistically stuck a halfpenny stamp.

The spot Buck took me to, leading the way direct as if he had known the ground all his life, was a full four hundred yards from the house. The last hundred wasn't easy going, either, through silent tangled woods sodden with the morning's drizzle, and the complete journey took us three and a half minutes, walking fast but not running.

"You're sure this is the place?" I asked: for I was profoundly doubtful if any man alive could have covered the distance on his own feet in a minute and forty seconds, the time which Buck had so specifically mentioned the night before in the dining-room.

"Dead sure," he replied. "See that oak over there with the hole in it? That's one of my regular calling-places. Cook used to make me up packets of sandwiches and three or four flasks of coffee, and I parked 'em in various spots, and that's one. I'd been here a good few minutes filling up when the fun began, as a matter of fact."

"And where before that—nearer the Grange?"

"No, farther off—away over there,"—and he pointed in the opposite direction.

"Good enough," I said. "Now,"—looking at my watch, "stand still and listen."

We did so, and both heard the three outside shots, faint but clear. The succeeding three we didn't notice at all, however, although Buck said the wind conditions were about the same, and Tony assured me later that he had duly discharged them. The appearance of the log—if not of the stamp—bore him out.

I dismissed Buck, after warning him to keep indoors that night, and before returning to the house went along to the lodge alone and had a word with the Judds. The chauffeur was taking life very easily today, sprawled in a wicker arm-chair by the kitchen range with his feet up and a paper on his knees, while his careworn wife washed up plates and dishes. In a cage by the window a disconsolate canary fluttered, and the room smelt of onions.

Mrs. Judd wiped her hands on her apron in traditional fashion when I said I would like to see them both together, and sat down nervously on a hard ancient sofa with her dull eyes fixed on her husband's ugly face. I asked for an account of their movements the previous evening, and both asserted that they had not stirred outside after seven o'clock until the police car arrived. I was watching them carefully, though, and got the impression that they were lying: perhaps more from the woman's manner than any-thing either actually said. Judd was apparently open enough, as if policemen held no terrors for him, but his wife was a shade too quick in muttering "Yes, that's right," in answer to his quick demanding glances for corroboration. I remembered that Tony had described him as not dull-witted, but I can't say I saw any great signs of intelli-gence: he would have appealed more to D. H. Lawrence than to Henry James. He addressed me as "Officer", and invited me to fill my pipe with his nauseating shag, but seemed not in the least offended when I declined.

I stayed only a short time, during which he confirmed that on Thursday morning he had made the partition shutting off the study window to Adair's instructions, and helped Hinkson to fix it up. He also declared that he had heard no shooting the night before, and Mrs. Judd agreed. As I walked back to the house I was wondering if the man's coarse masculinity would ever have been likely to appeal to Lina, and decided that it might have done if she were ut-terly devoid of taste.

XVII

"EVERYTHING satisfactory?" asked Tony, when we were alone in the study.

"Not by any means," I said, to his surprise.

"No? What isn't?"

"Buck's story. Can a man in ordinary clothes and a rain-coat possibly run nearly quarter of a mile, at night and part of it through woods, in a minute and forty seconds? Leaving out his improbable preciseness."

"Well, Jesse Owens might."

"And Buck?"

"No fear—his legs aren't long enough. Is that the time he told you? Then he made a mistake."

"Possibly—and possibly not, of the kind you mean. He was very definite about it last night—said he'd timed himself and thought it was pretty good going."

"Then how do you account for the—discrepancy?"

"An error of judgment: he didn't realize how long he would have taken to get to the house from where he decided to say he was. Distances between places always seem shorter if you know the ground, as he does, and consequently you tend to underestimate time."

"And the inference?"

"Primarily, that he wasn't where he told me, but I dare say you could have done that for yourself."

"I should hope so. I meant, what do you make of the discovery that he's been spinning fairy-tales?"

"Goodness knows. All we can be sure of is that we've one egg in our basket so far."

For the next hour we made further experiments with the pistol and revolver, fortunately having a sufficiency of ammunition. Hinkson's I offered to pay for, but he wouldn't hear of it, saying that he was quite as keen to settle the problem of Adair's death as I must be myself. He would have joined us at the slightest invitation, I think, but wasn't given the chance.

The results we arrived at, of any importance, may he summarized as follows.

(1): Shots fired from the little lawn could be heard from Andrews', Adair's, and Hinkson's bedrooms, but from none of the others. They were clearly audible everywhere downstairs, though.

(2): Shots fired from the study, with door and partition closed, were noticed from the kitchen and the communal lounge, and—surprisingly—from two bedrooms, Montague's and Hinkson's. Why from two only I don't pretend to understand: I merely record it as a fact.

"Then it seems," said Tony, "that Andrews and the cook and Montague and Hinkson all ought to have heard Adair killing himself: yet they all say they didn't. Any solution?"

"So you're on the side of the suicides, are you?" I remarked.

"Yes, 'fraid so. Deep thought has convinced me that murder was impossible. But answer my question."

"Certainly: they were all in league together not to hear."

"Oh, piffle—I don't believe that for a minute."

"Nor do I, really. Then have another answer. For some reason, not necessarily anything to do with Adair's death, none of those four persons was where he or she is supposed to have been at the time."

"Ah, that sounds better—only it probably means four different reasons. But what about Andrews and the cook: where else could they possibly have been?"

In reply I explained about the wireless set, and we determined by another experiment that their story would pass.

"Then that leaves two," said my friend. "And neither of them had a wireless."

"But Hinkson says he must have been asleep in the tree, don't forget."

"Oh yes, so he does. Yet he didn't seem to think he wouldn't have heard, did he?"

"No, and I don't think so either, so perhaps he wasn't in the tree at all."

"But surely he'd hardly draw attention to himself like that, would he?"

"Like what?"

"By saying he really thinks he would have heard. Wouldn't his line be 'Yes, I sleep like a corpse—I doubt if I'd have noticed a brass band'?"

"I don't fancy so: wouldn't that be even more suspicious? No, in my opinion the line he did take would be the only sensible one, whether he was where he says, or busy murdering Adair."

"And performing a miracle in getting out of the room again. Unless he was in there all the time, of course, but knows how to make himself invisible. You'll probably find he's a direct descendant of Siegfried, and inherited the Nibelung's net. And Montague?"

"I don't know—it'll bear thinking about. I shall be tackling him some time, though, because he's already told me one pretty obvious lie."

"Really? What was that?"

I explained, and he agreed with my view.

"Now, before, we leave the subject for the time being, tell me something else," I said. "If you hadn't had an opportunity of finding out, and you were asked from which bedroom, Hinkson's, Lina's or Tilly's, a shot from the study was likeliest not to be heard, which would you pick?"

He thought a moment, and then answered as I had expected.

"Montague's," he declared. "Montague's, without doubt. I suppose there's no chance that a silencer was used?" he added, so casually that I knew he believed he had had a brilliant idea.

"None, I'm afraid," I told him. "The muzzle of the pistol was moist when I found it, and I think I know saliva when I see the stuff, and that would effectively rule out the use of a silencer."

"Why? Don't they work in the damp?"

"I don't know, but you certainly can't fit one to a pistol, and then suck the end of the barrel."

"Oh well, if you say so I suppose it's all right. I know nothing about the things myself."

"No, I thought not. Any more brainwaves?"

"Well, there *is* another point, to do with Buck. If he wasn't where he took you just now, how could he tell that the shots Hinkson fired would have been audible from there?"

But that too was easy.

"By experience," I said. "The man probably knows practically everything about firearms and how far their noise carries—a lot more than he knows about how long it takes to get from A to B by running there."

I then disclosed to him Buck's true identity, and was rewarded with adequate astonishment; not unmixed, I fancied, with regret. Tony seemed to have taken quite a liking to the little ruffian, and I admit that in his position I might have done the same. We left the study soon afterwards, but not before I had examined the position of the books on the shelf containing the Macaulay, and satisfied myself that nothing had been touched. Tony eyed me curiously, but I deferred telling him of my morning's discovery until later, when we should have time to discuss it.

And here, I think, it will be as well if I give a brief description of the Long Corridor, and the disposition of the rooms that led off it. From the corner of the right-angled bend which hid it from the main hall its length was forty-two feet. On the left-hand side, going away from the hall, were three rooms: in order, the panelled room mentioned in Boon's diary, Adair's study, and an empty room of smaller size. On the opposite side were two only, both vacant. They overlooked the drive, and seemed to have been swept fairly recently. When I asked Andrews why this should be, although the panelled room was still in the chaotic state to which the treasure-searchers had reduced it, he smiled, and said that Lina had instructed him to find out if either was suitable for dancing in.

"But I doubt if they'll be used for that now," he added: though I gathered from his tone that he wouldn't put even that past Lina.

The time was now about a quarter past four, the sky dull, rain imminent, and dusk but a few minutes away. While I could still see I decided to inspect the beech trees more closely than I had been able to do so far. There were three, two on the left as one looked out of the study window and one on the right. In this latter Hinkson had sat, and perhaps dozed, if he were telling the truth. That he had indeed climbed it seemed evident from the smears on the bark. Yet when I went across to the other two I had a surprise: each showed traces of having been similarly dealt with at no far distant time. I also examined the ground under the study window, but it was free from footprints.

Indoors again, I sought out Hinkson in his room, where he was writing letters as he had threatened earlier, and asked if he had tried to climb more than the one tree. He said that he hadn't, and professed himself unable to offer any explanation of the marks on the others. He then made an odd remark.

"A bit awkward, that," he said. "I must think it out."

But when I pressed him to tell me what he meant he passed it off with a smile and a shrug. He assented readily to my suggestion that we should set about the desk in the study; but at the end of half an hour I was forced to the conclusion that it contained nothing of real interest. The chief discoveries were an unopened bottle of whisky and a clean tumbler, a cheque-book, a passport, a small glass percolator, and about half a pound of coffee in a vacuum tin. By the smell it was very good coffee, and Hinkson said that it was some Adair had sent to him regularly from South America.

"Did he have it often?" I asked.

"Fairly—always after dinner if he was working, and sometimes after lunch too. It costs a fabulous price, I believe—perhaps that's why he never used to offer me any. All I got was the smell."

Apart from the things I have mentioned we found nothing worth talking about, except a half-full box of cartridges to fit the Webley: no will, no interesting correspondence, and none of the translation of the parchment. I then asked Hinkson if his employer had kept a diary.

"I don't know. He had one of those rubbishy little pocket-book affairs, but I don't think he wrote in it. Not when I was about, anyway."

"Yes, I saw the thing," I said. "It wasn't any help. And apparently he didn't keep a waste-paper basket either."

"No—everything went on the fire. That's what's so queer, you know. He must have burnt every single line of the work he'd been doing the last few days, and he couldn't have done that in five minutes. And why do it at all? It beats me. He was so perky, too, all yesterday, even last thing—not a hint he was going to do himself in. Look here, Inspector, tell me what you really think. Are you absolutely certain he committed suicide?"

"Well," I answered, "I haven't found enough against the idea to make me consider murder seriously. I admit he seems to have behaved a bit queerly—destroying all his stuff, and so on: but people who kill themselves can't be expected to act normally. And anyway, plain facts point to the conclusion that murder's out of the question. Or don't you agree?"

"Yes, I suppose I must," he muttered, after some hesitation. "If a room's bolted and padlocked all round, and nobody can leave it like that from outside, and nobody can get up or down the chimney, and there aren't any hidden trap-doors, and you break in and find a man dead who's been shot at close range, and the gun's there and everything, then suicide simply shrieks at you. On the other hand, I hate being shrieked at, and I can't help feeling we oughtn't to forget the ancient and honourable profession of conjurer. Did you ever read a detective story by John Dickson Carr called *The Hollow Man?*"

"Yes," I said. "He gives a pretty good selection of ways of getting into and out of hermetically sealed rooms, if I remember."

"Exactly: mayn't one of them have been applied in the present instance?"

"In my opinion, no. It's absolutely certain that nobody could have played tricks with that padlock. It can only be closed with the key, and there's simply no way of doing that from the other side of the partition, and it definitely was closed. That rules the window out: or out of a conjurer's domain. To get anywhere you'd have to introduce real magic, and as far as I'm concerned magic doesn't happen nowadays. I believe Pharaoh kept one or two good men, but they're not in business any longer. Do you agree so far?"

"About the partition, yes: but I was thinking rather of the door."

He got up, walked over to it, and then turned back with a smile.

"May I touch?" he asked.

"Certainly."

"Thanks: only one gets so used to revising the formula, now everybody's crime-conscious. You know—'And I must warn you that anything you touch without gloves may be taken down and used as evidence.' "

He fiddled about with the lock and bolts for a few minutes, and then shook his head reluctantly.

"The wretched thing overlaps the door-post," he said. "Likewise the cross-beam at the top. I hadn't noticed properly before, but it rather rules out all the string methods of faking. Unless anything could have been done with that gap at the bottom—about an inch and a half, isn't it? He was always complaining of the draught, but never did anything about it. But that wouldn't be possible, either. And the hinges are this side, so nobody could fool about with them from the passage. Ah, but what about the partition ones?"

He hurried across, then shook his head again.

"They aren't on the window side," he murmured: a fact I had long ago noted.

"And yet I can't say I'm satisfied," he went on. "I don't *want* to be satisfied."

"Why not?" I demanded, a trifle crudely. "Why should it matter so much to you how the man passed on?"

He opened his eyes a little at that.

"Oh, not from any sentimental or high moral motives," he assured me. "If you won't hold it against me, I can't say I think the world's suffered much loss. My trouble is that I hate being beaten like this; or, more accurately, I hate not knowing whether I'm a fool to be taken in or a fool to imagine I may have been. Either way it comes to the same thing, vanity: I'll say it first to save you the bother. Still, I can't help the way my mind works."

"No, quite," I agreed: yet I was considerably more puzzled by Hinkson's attitude than I hope I showed. I couldn't understand why he should be so anxious to impress me, if that was his intention; or, if not, why he didn't keep his genuine doubts to himself for the moment, until he should have some practical suggestions to make. He didn't strike me as at all likely to be the garrulous unself-conscious type who habitually did his thinking aloud. I decided to spring a question at him.

"Why were you being so friendly with Buck after lunch?" I asked. "I thought you didn't like the fellow."

He riposted without hesitation, and I realized that I had been right. There was no wool in Hinkson's head: on his mettle he would be a cool and dangerous adversary with remarkable control of his wits.

"What made you think that?" he enquired, with the smile that was fast becoming familiar to me.

"Oh, general impressions, and something Tony said, though I forget what exactly."

"Well, you're quite right—I don't like him. That was why I was being friendly."

"I don't like parsnips," I said: "but that makes me avoid them, not eat them."

He gathered my meaning without comment on my way of putting it, and nodded.

"But it doesn't matter a rap about the parsnips' point of view: whether they know you hate them or whether they don't. It's rather different with Buck, though. You see, if there's been any trickery, then there's no question about who did it: he did. At least, no question in my mind, perhaps I ought to say. I never have trusted the man, and what's more, I don't believe Adair did either, very far. Still, I haven't a shred of real evidence, so I mustn't talk too freely. Just for the moment, even after looking at the door, I'm going to assume that things mayn't be quite what they seem, and accordingly I shall do my best not to let Buck think I've a notion that anything might be wrong.

"As it happens, I've always been a bit on the formal side with him—had to, or he'd have been all over me. Of course, I'm a bit sorry about that now, and all I'm hoping is that he won't think my change of attitude peculiar."

I laughed: I couldn't help it, remembering what Buck himself had said about Hinkson.

"He will if he's done anything to be suspected of," I said. "He'll be watching you closer than a nervous old lady watches a cow."

His expression was one of dismay.

"Oh lord, do you think so? Then what the devil had I better do?"

But I shook my head.

"I'm not up to these subtleties of conduct," I told him. "Why not write a post-card to Napoleon, c/o the *Invalides?*"

XVIII

ALONE in the study again, I first packed up the coffee and bottle of whisky for future analysis: a forlorn hope which I may as well say now came to absolutely nothing, since both proved to be perfectly pure. Then I sat down for a moment to think. Because of the small-pox scare I saw that I should have to forgo expert advice on the question of secret passages: but after looking round I felt fairly sure that I ought to be able to settle the matter myself. I got Tony and the constable—who had been taking a well-earned sleep—to help me, and long before an hour was up we were all convinced that the only ways out of the study were by the window or the door. In passing, I have never at any subsequent time had cause to change this opinion.

The constable's name was Johnson, by the way, and something in his appearance and athletic walk made me ask if he had ever done any running.

"Yes sir," he answered, "quite a bit. I've turned out for the county harriers before now."

"Ah, but I didn't mean long-distance stuff. Can you sprint at all? Could you do a quarter in under a minute?"

"Fifty-two seconds dead, last Whit Monday."

"Splendid. Would you object to doing one in uniform?"

"No sir,"—a puzzled expression on his not unhandsome face.

"And in the dark? I want to test the truth of some body's statement, and you seem to be just the person I'm looking for. I'll try to arrange it tonight some time, only I don't want an audience."

"Are you sure you'll be able to find Buck's sandwich dump again?" asked Tony, as we made our way to Adair's bedroom.

"Yes, I think so."

"Good—then if the copper can't do it in a minute and forty seconds, it's a pretty safe bet that Buck couldn't have. What are we supposed to be doing up here?"

"I don't really know," I said, gazing round the room: I had already been in it, but not to make any kind of search. The furniture was of the barest: bed, chest of drawers, one cane chair, and a small bamboo table which held a book or two and a half-smoked pipe. There was also a built-in cupboard next to the fireplace; the door was partly open, and one glance showed me that I had not come up in vain. On the floor was an old black safe, of the kind called portable: that is, weighing perhaps three-quarters of a hundredweight.

I felt in my pocket for Adair's bunch of keys, soon found the appropriate one, and inside we made the discovery I had been hoping for since the moment I saw it: the dead man's will, neatly rolled into a cardboard tube, tied up with pink ribbon, and sealed. Without hesitation I extracted it, and we examined the document together.

It was dated November 9th, 1936, less than a month after Tilly's twenty-first birthday, and for a will was surprisingly short and simple. To his daughter Adair had left £1000 only, "since she was adequately provided for by her late mother"; and free of duty, "as that has never been her strong point." Montague received £5000 and any books he cared to take, and the residue went to Lina without stipulations. Accompanying this last bequest was a somewhat flowery tribute to her "unfailing cheerfulness and companionship', and a reference to her father as "a man with whom I shared many a pleasant adventure in my youth."

"So!" said Tony. "The word 'motive' may now be written three times in capital letters. A thousand pounds wouldn't be much different from the Bank of England to Tilly, nor five thousand to Montague, if what he told me is true, and I know Lina doesn't despise easy money."

"How do you know that?" I asked.

"By my fiver that I haven't seen since it was taken off me in broad daylight. I shall send her in a bill. And I haven't told you what part of it went on, have I?"

He then did so, and asked if I agreed with what he had just said about motive.

"I don't know," I answered. "If this was a case of murder, it wasn't done all in a moment: it was carefully planned. I'll give you reasons for saying that tonight—we'll have a talk."

"Good," he remarked. "I thought it was about time you opened your mouth."

"Or my heart, perhaps you mean. Well, to go on with what I was saying: if murder, then premeditated murder; if premeditated murder, here are a couple of pertinent questions. Suppose Montague did it, and his motive was money: then would he go and tell you beforehand, and me afterwards, that he came down here solely for what he could get? Again, suppose Tilly did it: then what about her outburst on the golf links?"

"Well, the answer's easy," said my friend, "and the same for both. Or for anybody else. The murderer relied entirely on what he or she considered the certainty that Adair's death would be taken for suicide. Eventually, that is, if not to begin with. Hence no amount of potential motive would make an atom of difference."

"In short, he doesn't mind being suspected?"

"Not in the least: his retort is that there's nothing to suspect, and I think you'd have a bit of a job to contradict him, you know."

"Maybe. But in Montague's case there's still the question of why he didn't hear the shot, if it was suicide, and the other point I mentioned this afternoon, too. Still, there may be nothing in the first: he was reading, and deeply engrossed, or asleep, but didn't and doesn't know it. We'll talk about all that presently. The main point at the moment is what to do about the will. Ought I to show it to Lina and Tilly, or pass it on to the lawyers, or what?"

"Oh, the lawyers, surely. Who are they?"

"Trimby & Evans of Essex Street. But in that case, nobody will have authority to tell Hinkson to get on with the parchment. I'd better get on the telephone, I think, even though it is Saturday evening, and I'll probably be nagged for opening the blessed thing at all. However, I can always say I thought it was a matric. certificate."

I duly returned to the study—the telephone, by the way, was on a small flap near the partition, the lead-in coming through the window: I should have mentioned the fact

before. Eventually, with the patient help of various ex-
changes, I managed to locate Trimby at his home; but
unfortunately the man disclaimed any knowledge of Adair
after 1934. In that year he had had a disagreement with the
firm in connection with a threatened slander action against
him, and terminated all dealings with them. Trimby assured
me that he hadn't the least idea about what solicitors his
former client might subsequently have employed, said he
didn't think the slander action had ever been proceeded
with, and reluctantly disclosed the name of the proposed
prosecutor: Roger Montague.

That was news, and startling news, and immediately I
pressed for the name of the firm who had communicated
with Adair on Montague's behalf. Trimby, however, either
couldn't or wouldn't tell me, and almost at once brought our
talk to an end.

If I wanted to pursue the matter without delay I should
have to tackle Montague direct, I realized. After some
thought I decided to leave him alone for the time being, and
instead took upon myself the responsibility of telling Lina
and Tilly about the will. The dead man's daughter received
the information affecting her quite calmly.

"A thousand pounds for me, and the rest to Lina?" she
repeated.

"More or less, except for one other bequest of no very
great consequence. At least, that might depend: can you
give me any clear idea of what your father's estate is likely
to be worth, Miss Adair? Excluding the possibility of a
treasure on the premises here, of course."

"But including the ruby?" she asked sharply, staring at
me steadily through her thick glasses. She was still in black,
sitting by the window in the smaller sitting-room, the one
for the girls. Her face looked less unattractive by lamplight,
I thought, but the way she had drawn her legs up under her
produced an unfortunate impression that she had had them
amputated.

"Yes, I suppose that would come in the residue," I said,
and was immediately sorry: I had not meant to rub in the
fact that Lina was getting such a disproportionately large
share.

"Provided it turns up again, of course," I added.

"Yes," she agreed, and sighed heavily.

"A thousand pounds," she repeated. "It might be worse, Inspector—at least I shan't starve for a year or two, and I can probably get a tutoring job or something like that. But you wanted to know how much there'll be altogether. It's hard to say, but Daddy told me once that he could always raise a hundred thousand on his investments and securities if he ever had to."

"I see—thank you. By the way, you didn't know that he'd stopped employing Trimby & Evans as his solicitors?"

"No. Then whom did he have after them?"

"I can't say, but his bankers may have been told."

I ought to have thought of that before, and at once telephoned Scotland Yard, asking for a man to be sent round first thing on Monday morning to the Lombard Street branch of the City & Provincial Bank.

Next I searched for Lina, and found her conversing in a low voice with Buck in the dining-room. Each held a glass of what looked like sherry, and neither seemed particularly pleased at my interruption.

"Can I have a word with you alone, Miss Hipple, please?" I requested.

"Oh, all right," she agreed, not very graciously; at which Buck at once got up, grinned at us familiarly, and slouched out.

It was evident at a glance that Lina wasn't going to pursue any policy of ostentatious grief for Adair. She had changed her black frock for an equally becoming one of a darkish blue, ornamented with chromium buckles, and her finger-nails were still enamelled a brilliant red.

"Yes?" she queried, lighting a cork-tipped cigarette.

From the way she looked up at me over the match flame I had the feeling that she thought she knew what I wanted to speak to her about, and this worried me a little. I should very much have liked to be sure of the subjects upon which she was prepared for interrogation. I told her about the will in a few words, and her change of manner bore out my suspicions: she was undeniably relieved. She made no attempt whatever to hide her gratification at finding she was Adair's principal heir, and almost clapped her hands with excitement.

"I say, I wasn't expecting all that!" she cried. "And Tilly only gets how much? A thousand? My God, won't she grind her teeth! Still, I mustn't be nasty about it, must I? There'll be enough for her to have a bit more than that, I dare say. Thanks terribly for telling me—have a cigarette?"

"Not at the moment, thank you," I said, doing my best to glare at her. She really was behaving disgustingly, and to my eye her face had lost all its former beauty, being now but a mask hiding a harpy. If I had been feeling at alt fanciful I might easily have imagined that her nails were claws red with her benefactor's blood. She was smiling, as if inviting me to rejoice with her, and I felt that I wanted nothing so much at the moment as to remove that smile.

"Well, I've heard of ill winds," I said, "but I never thought I'd see anybody so obviously glad to step into a dead man's shoes. If Major Adair's death weren't so plainly a case of suicide, Miss Hipple, I'd be inclined to put the old principle of *Cui bono?* into practice, and arrest you on the spot."

"What d'you mean?" she gasped, her delight a thing of the past. "*Cui bono?* What's that?"

"As far as sudden death goes, it means in plain English 'Who stood to gain most?'. The idea being, of course, that people who want a thing badly enough will sometimes do even murder for it. However, don't let me frighten you."

"Oh, I'm not frightened!" she said: and indeed she looked more angry than anything, though I wouldn't have taken an oath that she wasn't just a bit afraid of something. "I think you're being confoundedly rude, though—what business is it of yours?"

"How you take the news that you've been left a fortune? None—so far. I thought you'd better be told, though. Still, you mustn't go building your castles too high. I can't guarantee that the Major didn't make a later will than the one I found, leaving everything to a cats' home."

"Oh!" she exclaimed. "Well, when shall I know definitely?"

"I can't possibly say—at present I haven't even found out who his solicitors were. Can you tell me?"

"No—can't Tilly?"

"Apparently not. However, you'll probably be fairly safe in assuming that this house will eventually be yours. Now, earlier today Mr. Hinkson suggested to me that it would be a good idea if he tried to translate the last part of the parchment clue to the treasure—he seems to think he might manage it. Since it's presumably yours, I'll tell him that you're the person to ask permission from, shall I?"

"Oh yes, rather—or I will. He can get on with it right away. How long did you say this blasted place is in quarantine for?"

"A fortnight. They'll be sure to adjourn the inquest until we can attend, though, if that's what you're thinking of."

"Inquest? But surely there'll have to—I mean, he'll be buried before then, won't he?"

"Naturally: but burial doesn't terminate a coroner's enquiry."

"Then what does?"

"The facts of the particular case. When he's satisfied that there was no question of foul play he'll deliver a verdict accordingly, or his jury will."

"But of course there wasn't any foul play—how could there have been?"

"I don't know—I simply can't imagine," I told her, and went away thoughtful. At about midday she had declared her inability to believe in suicide: yet here she was at quarter to seven declaring with equal vehemence that suicide was the only reasonable solution. Such a complete *volte-face* seemed to require attention, even though she might have had her cue from me.

XIX

ALTHOUGH Lina had said that she would speak to Hinkson about the parchment herself, I made a point of seeing him the moment I had left her. I had to tell him that she was probably Adair's principal beneficiary, of course: he didn't seem particularly surprised, and said he was glad he was free to get on with the clue.

"I ought to be able to manage those last two paragraphs in a week at most," he assured me: and then abruptly changed the subject.

"Remember you wanted to know why all the trees seemed to have been climbed recently?" he asked. "Well, I think I know why. What Adair said to me about them was this. 'You'd better take the right-hand one—it's the easiest to climb.' "

He paused expectantly then, but I remained silent, which I think disappointed him.

"You see the point?" he was forced to add at last. "How could he possibly know which was the easiest unless he'd tried them all?"

I wasn't impressed by his reasoning, and said so.

"You might almost as well ask how anybody could tell which was the tallest till he'd been over them with a footrule," I remarked. "It doesn't take a lot of judgment to know whether a tree's going to be easy or difficult."

"Oh, have it your own way," he replied shortly. "I bet I'm right, though—Adair was that sort of man."

"Then it shouldn't be difficult to determine," I said. "Or was he very particular about his clothes?"

"Clothes? Oh, you mean there'd be marks and things on them. Yes—good idea. I don't suppose he ever used a brush from one year's end to another. But the ones he was wearing last time I saw him looked all right."

He offered to show me the way to the bedroom, not apparently realizing I must already have been up there to have found the will, and in the wardrobe, jumbled in a corner, we came on a dark plus-four suit bearing obvious traces of contact with bark. Hinkson was as pleased about the verification of his views as if he had located the treasure itself. He struck me forcibly as being a conceited young man: perhaps a very conceited one.

Dinner was at 7.30 that evening, and nobody changed. It was almost as silent a meal as lunch had been, and afterwards the other three drifted away without staying for coffee. At quarter past eight Judd brought me up a sealed envelope which he said had been delivered at the lodge a few minutes earlier. It contained the police-surgeon's preliminary report on the body, and the opinions he expressed were briefly these.

Adair had been killed by the entry into his brain of a bullet from a .25 automatic pistol, fired with the muzzle in the mouth. The bullet had lodged in the skull and was now being examined to determine if it had come from the Webley. Death, in the doctor's view, had been instantaneous, and had taken place between 9.00 and 10.00 the previous evening—nearer than that he wasn't inclined to go. Everything—nature of wound and condition of body and pistol—pointed to the probability that the shot had been fired by the dead man himself, and he had noticed nothing to

make him think that it was other than a clear case of suicide.

"So much for that," I said, passing the report to Tony. "If
it wasn't suicide, then somebody's gone to a deuce of a lot
of trouble to make us believe otherwise. Still, that's only to
be expected—there wouldn't be any sense in botching a job
like that. We'll have our talk in a minute, but first I want to
do a bit of tree-climbing."

"What on earth for?" demanded my friend, "Fed up with
that suit? I thought it was rather nice."

"It is, and I don't intend to wear it. As for why, I'd like to
find out if anyone outside the room could possibly have
seen what was going on inside. Come along to the study,
and I'll try to arrange things as they were when I broke in
last night."

We soon prepared the room, Tony playing the part of the
body. Johnson I sent off for a break, telling him that I
wanted to make use of his running powers shortly. There's
no need to elaborate what took place next: but, in case it
hasn't been made quite clear, I will repeat that the beech
from which Hinkson said he had fired stood by itself at one
end of the small lawn, and the two others at the opposite
end.

After one or two false starts I managed to get into each
tree in turn, but only from the smaller one of the pair was it
possible to see into the study at all, through the chink at the
top of the wooden partition shutting off the window. I
levered myself about on the bare branches, and tore several holes in the ancient pair of flannels I had put on, but
however much I twisted myself I never managed to see the
desk, or Tony sitting at it. The normal limit of my vision was
the bottom half of the door into the Long Corridor, and a
narrow strip of carpet, but by craning my neck I could just
include the lock and handle as well, though only for a
second at a time. As can be imagined, I was soon driven to
the conclusion that it wouldn't have been worth anybody's
while to sit long in the tree in order to learn what might be
going on in the study by the use of eyes alone.

The result of testing the probable untruth of Buck's
statement was more satisfactory, though. I induced
Johnson to make three trips, going at full speed on each
occasion, and his times, in order, were 2 minutes 6 seconds,
1 minute 59 seconds, and 2 minutes 1 second. When I

asked him if he thought anybody could have done the distance in a minute and forty seconds, however good a runner and however well he might know the ground, he answered me with an emphatic no.

"He couldn't go all out through the wood part—nobody could," he declared. "If you were to put Roberts or Brown or any of them cracks on the job, and give 'em the right kit and a week to practise the course, I don't believe you'd get anything under a minute and three-quarters, Inspector."

I thanked him for his help, cautioned him against telling anyone what he had been doing, returned indoors, locked up the study for the night—the door had been mended—after inspecting the Macaulay again and finding it still untouched, and made my way to Tony's room. I expected to find him alone, but heard voices as I approached the door, and there he was standing by the fire, Hinkson on the bed and Lina reclining gracefully in the arm-chair.

"These people want to see you about something, Ted," he explained briefly. "I'd better clear out, shall I?"

"Oh no, please don't," said Hinkson quickly. "Not the slightest need, is there, Lina? He ought to be here, really."

"Yes," agreed the girl, though without much warmth. She looked less at her ease than usual, I thought, and was wearing black again, though not the same frock as she had had on that morning. Whatever else might be obscure, it seemed certain that Adair's adopted daughter had received an ample dress allowance. The secretary was in his same smart suit, and both eyed my dishevelled appearance rather pointedly. This annoyed me, because I didn't want them or anyone else to know I had been climbing trees. Accordingly I tried to put them off with a lie, saying that I had been exploring the grounds, but had seen nothing out of the ordinary.

"Why, what did you expect to see?" drawled the girl, and I fancied that she was deliberately prolonging the conversation, edging away from the subject she and Hinkson had apparently come to discuss. When I find people in that condition I try to come to the point quickly, in case they haven't quite made up their minds how much they're prepared to say. Often it's possible to hustle them into disclosing more than they meant to.

"I thought there might be somebody prowling," I answered. "But you needn't worry about sleeping soundly—

there'll be men on duty in the woods all night. Now, what are we going to talk about: the ruby, the treasure, or the Major's death?"

Purposely I rather barked the question, and addressed her, not Hinkson: a glance had shown me that he would be cool enough. Nevertheless, he it was who answered.

"We've come to tidy up our position," he said, every word clear. "You see, Inspector, Lina and I haven't been telling you the whole truth."

"No? You amaze me."

"I wonder," he murmured. "I wonder if I do."

"Well, perhaps not yourself," I conceded—it wouldn't do to be caught pretending to be too dense. "There was undoubtedly a shot fired in the study between 9.00 and 10.00, and if you were in the tree I can't help thinking you ought to have heard it."

"Very nicely put," he said with a smile. "Yes, I expect I should have done."

"Good—then where were you?"

He smiled again, and crossed one elegant leg over the other, entirely composed.

"Can I explain in my own words? Thanks—then this is what really happened last night. In consequence of a remark made by Major Adair on Thursday afternoon, Lina and I were getting a bit restive. Apropos of nothing in particular that I could see, and right in the middle of talking, about something else, he suddenly told me he hoped the ruby was all right, and that he ought to have locked it up. Luckily he didn't pursue the subject, but it came as a blow to find it so obviously in his thoughts, I can assure you. It seemed quite possible that he wouldn't hold out even till you got here, Inspector, and if he asked for the wretched thing before it turned up, we were sunk. Anyway, that was what I believed then.

"I spent quite a bit of time after that thinking, and the result was that I got out of playing bridge that night—remember, Tony? Not to beat about the bush, it was my intention to have a good look round Tilly's room: which I did, with Lina's help.

"Put baldly like that it sounds a pretty mean action, I dare say, but neither of us was quite satisfied she knew as little as she said; or anyway, I wasn't. We didn't find anything, though, needless to remark."

"Why needless?" I asked quickly, and once more he smiled.

"Now don't rush me, please—I'll tell you everything in due course. Well, I'd meant to go through Mr. Montague's room as well, but Tilly's was so untidy, and took us so long, that I didn't feel like risking it that night. You see, I thought the bridge would be stopping ages earlier than it did—I'd heard Adair say he meant to be in bed by half past ten.

"I didn't naturally have another opportunity to tackle Montague's room till last night, and the problem was how to get him out of it. Now, I don't want to tell tales out of school or anything, but Lina happened to know that Montague wasn't, well, indifferent to her, and that if she invited him along to her room for half an hour's chat he'd accept like the proverbial shot. I hope I needn't be coarse: again it was a mean trick, I suppose, as Lina doesn't care much for the man, and hadn't the slightest intention of letting him do more than hold her hand, but it seemed to us a good way of keeping the fellow occupied for a while.

"Well, directly after dinner she carried him off, I went back to the study. Not to work, though: I was going to get Adair to let me off for an hour even if I had to sham dead to do it. But then he trotted out his instructions about tree-climbing, and that just suited me, of course. What didn't suit was another instruction I received at the same time—one I haven't told you about. I was to send Lina along, with the ruby!"

He broke off, looking at me with his head a little sideways, as though to see if I appreciated the awkwardness of the position. I nodded wisely, and he seemed satisfied.

"I had no choice," he went on. "I tapped at her door and delivered the message, while Montague pretended to be looking at the pictures on the wall. It made her gasp, of course, and she turned all sorts of colours, but luckily kept her head. 'You stay here, Roger,' she told him—'I won't be long. Now don't go away, mind—you've promised to read my hand for me.' He smirked a bit at that, and assured her he wouldn't dream of budging, and she cleared off downstairs. I stayed behind a minute or two, and then made some excuse, and went along the passage to his room, two doors away.

"I reckoned I should be pretty safe for a while: Montague would be waiting patiently for Lina, and the old man—I mean, Major Adair—wouldn't be likely to stick his head out of the window to see if I really was in his precious tree. He'd have too much to worry about when he learnt the ruby was missing. By the way, Lina and I had arranged that she was to keep Montague with her till at least 9.45, so I felt all right till then—I didn't imagine she'd be in the study very long.

"Well, the rest of my part is easy to tell. I went up to my own room dead on quarter to, not having found a thing, collected my coat, nipped downstairs and into the garden, and scrambled into my tree. I stayed there till I saw that fellow creeping round about twenty past ten, and fired at him, as you know. I didn't go to sleep, and I didn't hear any sound from the study."

"You're sure of that?" I asked.

"Absolutely—I'm certain Adair must have already shot himself by ten minutes to."

"Which means that you've given up the murder theory?"

"Of course. The only reason I ever pretended I believed it was because I didn't want to have to tell you all this—about snooping round other people's rooms and so on. I had a devil of a job to keep a straight face this afternoon when I was messing about with the door and saying I was worried sick in case anybody'd been doing conjuring tricks. And when I came to think about it I was pretty sure anyway that you wouldn't have believed my story last night.

"Well, there you are, Inspector, and I hope you won't be too annoyed."

But I was, and told him what I thought of his deception pretty plainly. He had the grace to look somewhat ashamed, and assured me he was sorry for any inconvenience he had caused.

"Yes—I think you should be," I said.

"No, wait a bit, wait a bit," broke in Lina suddenly, causing me to turn in her direction. She was twisting her hands in her lap, and then abruptly rose, facing Hinkson.

"I'm going to tell him, Dick," she exclaimed in a low voice. "Yes, I am—I'll feel better, and there won't be any trouble."

"No, don't," he urged, but she nodded emphatically.

"Yes, I'm going to," she repeated, and then looked at me, her face strained and her lips inclined to tremble.

"Now listen to what I've got to say," she requested. "Everything Dick's said is true—except the last. It wasn't because he didn't want you to think badly of him that he pretended he thought Uncle was murdered—it was so you wouldn't know what a little beast I've been. You see—you see, I stole that bloody ruby myself."

"Corks!" I blurted out, in genuine astonishment: not so much because she was a thief—it will be remembered that I had already considered the possibility—as at her voluntary disclosure of the fact. To Tony also it came as a surprise, judging by his expression, in spite of what he had said over the telephone on Wednesday: but not to Hinkson. Plainly he already knew, and he now took Lina's hand in his and pressed it gently, as if to give her confidence.

"Now let's get this business straight," I said. "I don't want to embarrass you unduly, Miss Hipple, but I shall have to ask some questions. First, where is the ruby now?"

"I don't know for sure," she answered miserably. For the present it seemed that all the stuffing was out of her: even the bosom of her close-fitting frock looked to have sagged a little from its former impudence. "You see, I was broke," she explained. "I'd overspent my allowance, and borrowed and borrowed, and altogether I was in a pretty filthy jam. And then the ruby turned up, and I just couldn't resist taking it. I sent the cursed thing off days ago to the—the person I'd borrowed from."

"A moneylender?"

"Yes,"—with a shamefaced nod. "*The Greater London Security Trust,* he calls himself, but he's really a nasty bald little Jew called Hyams."

"Yes, I've met him," I said. "What did he allow you?"

"I don't know yet—I haven't heard except a note of acknowledgement. I expect he's getting it valued; but it's bound to settle everything—I don't owe him more than three hundred."

"And I'm to understand that all this business of getting me down here to look for it was sheer nonsense?"

"Yes,"—weakly.

"Then why do it?"

"Well—because I wanted to pretend I didn't know any more about it than anybody else."

"Perhaps I'd better explain," interrupted Hinkson. "When I approached Mr. Purdon I hadn't the slightest idea what had really happened. Lina didn't tell me till yesterday evening."

"Oh, I see: then you honestly thought my coming down here might do some good?"

"Yes: it was my suggestion to Lina, and she naturally supported it."

"Of course," agreed the girl. "What else could I do?"

I suppressed the obvious rejoinder, and asked what had made her tell Hinkson the truth after so much delay.

At that her face became more wretched than ever, and she glanced up with a harassed look first at him and then at me. (She had sat down again a minute or two earlier.)

"Because—because we happened to be quite fond of one another," she said at last.

"And still are," added Hinkson, quickly and pointedly: at which they exchanged a smile.

"Very well," I said: "now tell me about what happened when you got to the study last night. At about five past nine you were down there, Montague was in your room waiting to tell your fortune, and Mr. Hinkson was in his room looking for the ruby, which he still believed to be genuinely missing. Is that right?"

"Yes. Well, I went in, and the first thing the Major said was 'Got that stone? Let's have it.' I tried to put him off with the same old yarn about not knowing what had happened to it, and he just blew up."

"But I thought he was very fond of you."

"Yes, so he was: but that only made things all the worse. You see, it's a long story, but a long time ago my father and he were partners abroad—somewhere in Africa, to do with a gold mine or something. That was before I was born, and my father ran off with all the profits. What ever made the Major adopt me I don't know, but he always said it was because he wanted to prove for himself that there was nothing in heredity. You know, doing the same things as your parents did because you've got it in your blood, and all that. He made me answer thousands of questions first, and other people too: whether I'd ever stolen money, or pinched jam out of the cupboard, or found things lying about and not told anyone. I didn't understand it all then, of course, but he explained about a year ago.

"Well, the minute I said I'd lost the ruby last night he told me flat to my face I was a damned little liar. 'I knew it!' he said. 'I've been waiting for this, and hoping to God it'd never happen, and it has. Like father like child—I spend years looking after you, and hundreds of pounds trying to make you a lady, and all I get for my pains is a crack on the jaw like this.'"

The last words came rather falteringly, and for a moment I thought she would break down; but she set her teeth, and went on in a thin hard voice.

"He said terrible things," she told us. "He was an old man, cursed with an ugly cretin for a real daughter and a pretty crook for an adopted one, and a houseful of vultures all waiting to dig in at his treasure. He said—he said we were all a lot of hell-hounds, and it made him sick to be alive in the same house with us. And he put all his papers straight in the fire, and swore at me, and said if I wanted the treasure I could find it myself, and then he suddenly got up and smacked my face hard and threw me out, and bolted the door.

"Oh, it's all horrible! He was nearly off his head—I could hear him muttering and talking to himself in there till I couldn't stand it any more, and went up to my room. I might just as well have shot him myself—I wouldn't be any more of a murderess."

I said nothing to that, giving her a minute or so in which to recover something of her composure. Then I pressed on with my questions, and if anyone thinks I was behaving callously I can only say that although I felt sorry for the girl I felt a deal sorrier for Adair and especially for Tilly, who had so nearly drowned herself over absolutely nothing.

At the end of another twenty minutes I had all the information she could give me. The part relating to her exact movements (and Hinkson's), with the relevant times, I shall reproduce later; in addition, Lina was positive upon the following points.

Adair did not threaten to take his life in so many words, but she felt convinced that he had in fact done so.

He had extracted from her an admission of her theft, but didn't ask what she had done with the ruby.

As far as she knew, no one could have overheard their conversation.

When she left the study, only the parchment remained upon the desk, in the way of papers: all the rest had gone into the fire.

The decanter was on the mantelpiece, and—she thought—was half full. (Hinkson confirmed that that was its usual place during the evenings, and said that it had been three- quarters full when he went to change for dinner. He didn't remember noticing it when he returned later.)

The partition was padlocked.

She had known nothing about a will previous to my telling her—not even that Adair had made one. She agreed that in his condition of resentment against her he might have been expected to bequeath his money differently before shooting himself.

She had not sat down during the interview, which lasted under ten minutes, nor touched any of the furniture.

When she got back to her room and found Montague still waiting she told him nothing of what had just happened, but remained chatting with him till it was safe to let him go. Then she complained of a headache, just before ten minutes to ten, and when she was alone lay on her bed thinking.

"And exactly when did you tell Mr. Hinkson the truth?" I asked, as her final question.

"Oh, not till early this morning—about three o'clock. I just couldn't sleep, and so I went along to his room and got it off my mind."

"And I advised her to say nothing unless she was certain she ought to," added the secretary. "It meant telling you an awful lot of lies, of course, and I'm sorry, but I'm afraid I'd do the same again."

"I hope it won't be necessary," I said. "By the way, have you started on the parchment yet?"

"No, not properly, but I've been looking through my notes again. And that reminds me—can I go into the study tomorrow? I shall want the tracing paper and dividers and things."

"Of course," I told him. "Work in there if you like."

"Oh, no thanks—I don't think I'd care about that, somehow": and he gave a shudder that might or might not have been an expression of honest repugnance. He then got up, but I had still another question for him before I let him go.

"It's all clear now but one thing," I said. "Why didn't you see Andrews and myself come round to look through the window?"

"Oh, but I did: what you mean is, why didn't I call out or something? Well, it's rather hard to say. To begin with I felt curious about who you were, and then I started wondering what in the world you were doing, and by the time I'd watched for a bit it was too late to make myself known. Can you understand? By delaying I'd worked myself into a thoroughly false position, besides feeling an utter fool squatting up there like a monkey on a stick, and so I preferred to keep quiet."

"Yes, I think I see," I said: and indeed, his explanation seemed to me to be quite understandable.

XX

WHEN we were alone again Tony gave expression to his feelings about Lina in very plain language.

"The nastiest little so-and-so I've ever met," he declared. "I take back nothing I've said or thought about her, and that's a lot, all told; and I was right about her and Hinkson, too. But things look a deal simpler now, don't you think? We've a good motive for suicide at last, and two fewer people who say it must have been murder, if that matters."

"And we also have a fat packet of questions for one Roger Montague," I added. "I'm not going to try to reckon where we stand till I've seen him. If he checks her story, well and good."

"Except over what he told me about definitely disliking her."

"Oh, that needn't worry you. He'd probably be ashamed for anyone to suspect he was attracted, and go out of. his way to pretend he hated the sight or her. I think perhaps you'd better not come along—he mightn't speak so freely in front of someone he knows."

Tony sighed.

"Well, don't be too long, or I shall go to bed. Anything you can suggest I do to pass the time?"

"Yes," I said, suddenly remembering that I hadn't yet told him about the paper between the pages of the Macaulay. He listened with interest, read it hurriedly, and whistled.

"We now have two motives for suicide," he declared, echoing my earlier thought. "It really does sound the most arrant nonsense—whatever arrant means. I never did know—do you?"

"Not now," I said. "Mind you've got a simple solution for me when I come back."

I found Montague in bed but still awake, an oil lamp on the window-sill beside him and *The Newgate Calendar* in his hand. He was wearing a soiled pink woollen dressing-gown instead of his previous brocaded one, and could say nothing articulate till he had adjusted his dental plate.

"Cheerful stuff, this," he then observed, tapping the skull and crossbones on the back of the book. "Ever read the history of the Rev. Thomas Hunter? Two small children caught him misbehaving with a servant girl, and told their parents, who employed him as their tutor, so in due time he cut the little darlings' throats with a penknife. In 1700, that was, and they suited the punishment to the crime—cut off his hand and then gibbeted him. But what a man! What a mind! I'd like to hear Freud's opinions."

"And I'd like to hear yours," I said. "About Lina Hipple, for choice."

He looked at me very hard, raised his eyebrows with a cynical smile, and nodded.

"Caught again!" he remarked, with an air of indifference; though I felt pretty sure that inwardly he was discomfited and furious. "You'd like more than a mere opinion, I don't doubt?"

"Yes please—an explanation as well."

"All right: you tell me what you think you know, and I'll fill in the gaps, or correct you if you're wrong."

I accepted the offer, and recounted some of the information I had just had from Lina, watching him closely as I spoke. He showed no visible signs of perturbation, however, and readily agreed that her story was substantially right.

"Yes, I went along to her room after dinner, about ten to nine, I suppose, and I stayed there an hour. Nothing indecent took place, though."

"No one has suggested that it did," I said.

"Dear, me! A case of *Qui s'excuse s'accuse,* I'm afraid. Then what made her tell you?"

"Well, call it a guilty conscience."

"Nonsense—she hasn't one."

"Then call it persistent bullying on my part. Now please answer this question. Did you notice anything at all out of the ordinary between the time you returned here, about ten minutes to ten, and the time you were informed by me of

Adair's death?" (It will be seen that my wording covered the possibility that he had seen something wrong with the state of his room as well as the matter of a shot.)

"Nothing whatever," he assured me earnestly. "I've only told you once—that I was here solidly from dinner."

"I see. Are you habitually a truthful man, Mr. Montague?"

"Why yes, I think so—as the world goes, you know."

"Ah, that's qualifying it a bit. Still, try to stick to your good habit, because I want to know your real opinion about Adair's death. Do you think it was suicide or murder?"

He smiled, and his lined face lost a little of its cynicism.

"No, I told you the truth there as well, Inspector. If it simply couldn't have been murder, I'll accept suicide under protest. If it could have been, with whatever amount of difficulty or improbability, then murder will get my vote. I repeat what I said before: in my view, Adair could never have plucked up enough courage to kill himself."

"That being your sole reason?"

"Yes. I've no suspicions against any particular person, if that's what you mean. Now tell me what you think."

"I don't—yet," I answered. "Or I won't, perhaps."

"I see: your judgment waits upon circumstance. You aren't trying to cast me for any sinister roles, I hope? Since my departure from good habits?"

"Well, of designs upon Lina, perhaps," I suggested gravely.

"Oh, but there I had an understandable motive. I may be old, but my blood's still thicker than water. But I wasn't referring to her."

"Then presumably you meant, do I suspect you of murder? It's rather a silly question, isn't it? Until I can find how it could possibly have been murder I'm not in a position to suspect anyone, whatever my inclinations to do so. In general, though, Mr. Montague, absence of known motive doesn't necessarily imply innocence. You agree?"

"Of course it doesn't. *Tot homicidia, tot rationes,* as Cicero might have said."

"Possibly: and motive is only a relative term, anyway, in spite of all the nonsense talked about it."

"Especially by reviewers of detective stories," he put in. "Reading some of them, you'd think that to make A want to kill B, it would at least be essential for B to have seduced A's wife, bamboozled him into bankruptcy, and blackmailed him heavily for a long time."

"Quite so," I agreed. "And yet in real life a murder-motive may seem the flimsiest possible thing to everyone but the murderer. Your Thomas Hunter, for instance, or Muller, who in 1864 killed a total stranger in a railway carriage for a gold watch and chain, or Scottie Mason in 1923, or even Browne and Kennedy: would any of them be considered credible by the pundits?"

"No fear: but then, pundits only judge by their own limited experience and mental processes. Naturally you must know a great deal more about criminal matters than I do, Inspector, but I have read a fair amount on the subject, and even written about it. Real crime, that is, not fiction. In my opinion, to want to kill a particular person, and to be capable of killing him, are two vastly different things, and I'd say that the greater the capability, the less need is there for universality in the motive. And this capability—what is it? I've never been able to make up my mind. Either abnormal self-control, or abnormal lack of it, I think."

"Corresponding roughly to premeditated and unpremeditated murders," I said. "But we're wandering a little from the point. I was saying that apparent lack of motive is no sure guarantee of innocence. Now, naturally the converse is equally true, but you'll agree it's decidedly more suspicious. In the present case I feel that I should need to meet three conditions before I could begin to consider any special person for the part of Adair's murderer. First, I must be confident the man was murdered. Second, this person must be unable to produce corroboration for his account of his movements during the *whole* of the period 9.00 to 10.00. Third, he must be able to be credited with at least one reason for being glad of Adair's funeral. Do you accept this?"

"Yes, certainly. I think you'd be perfectly justified in pointing your finger at him."

"And would I still be justified if I couldn't fulfill the first condition, but had to rely on mere opinion?"

"Not in general, perhaps, but I won't object here, because the opinion's my own, isn't it?"

"Yes—backed up by Tilly's, incidentally."

"And the others?" he asked sharply. "Lina and Hinkson? They both told me they thought the same."

"And me: but they've now disclosed reasons for switching over to the idea of suicide."

"Really? Adequate reasons?"

"I think so: for them, that is."

"Ah! One couldn't hope for more precision?"

"I'm afraid one couldn't, at present."

He nodded slowly at that, fingering his nose thoughtfully and regarding me with one eye almost closed.

"All right," he said: "who is it? This special person who could have had a motive and can't account for all his movements? Or is it a she?"

"No," I told him: "it's yourself, Mr. Montague."

His surprise, to my own, struck me as being quite unfeigned. He pursed his mouth tightly, opened his pale eyes wider, and kept his head very still against the propped-up pillows.

"Myself, eh?" he remarked at last, in a controlled voice. "That's using my own words against me with a vengeance, I'm damned if it isn't! You can't prove it was murder, so you get me to say I think it was, and then you practically ask me how I did it."

"Now, it isn't quite as unfair as that," I objected. "I did say that Tilly backed you up, and you must allow me sense enough to see that if you're a murderer then you're a very clever one, and quite capable of trying to bluff me by swearing it couldn't be suicide."

He laughed mirthlessly through his teeth at that.

"You're a queer chap," he said. "I honestly don't know if I'm supposed to take you seriously or not. Still, I'll be generous, and let what Tilly thinks fulfill the first condition. Now explain why you pick on me. What was my motive? Where did the opportunity come in?"

"Well, that's obvious enough, surely. There's only your word for what you were doing between five past nine and quarter past: you can't prove it."

"But why should I? Or, what's more to the point, how the devil could I be expected to? For instance, could you even prove you're in here now? You know you are, and I know you are, but how would you set about convincing somebody in London? By bringing me forward as a witness? And if I were to deny you'd ever been inside the room, would that prove you're not here now?"

I laughed.

"I dare say the question of our respective reputations for veracity would come into it then," I remarked. "But I don't expect you to be able to prove where you were between 9.05 and 9.15 last night: my point is merely that you can't. If I choose to disbelieve what you tell me, then for all I know, or you can show, you may have been downstairs shooting Adair. Always providing you've learnt how to get out of a locked room, of course."

"But Lina was supposed to be in the study during that particular period, don't forget. Presumably she's in it too?"

"Perhaps," I agreed lightly. "However, the question of opportunity doesn't really worry me much at the moment: I'd rather get on to motive."

"Very well: and I hope you won't be quite so feeble there. Firmly denying that I did murder Adair, or could have done, I ask you to tell me why I should ever have wanted to."

"Well, I happen to be pretty sure that at one time you bore him a grudge," I said. "You were going to bring an action against him for slander, weren't you?"

Again I had the feeling that he was really surprised.

"Now how the dickens did you get to hear about that?" he demanded.

"By telephone," I answered, truly. "I admit I haven't the details at my finger-tips, but I don't doubt I can get hold of them if need be. I thought perhaps you might prefer to explain things yourself, though, so I'm giving you the chance."

"Thanks,"—with pointed emphasis. "Yes, I suppose it would be better if you heard the correct version. But you're completely on the wrong track—there's no question of my having borne him a grudge. The facts are very simple. I first met Adair in 1897, and we kept in fairly close contact till the War. That took us different ways eventually, him to Whitehall, or France at farthest, and myself to France and then to Palestine. I didn't see him between 1917 and 1929, and that's a pretty long pause in a friendship. When we did run across one another again, the old footing had somehow gone, and for a time we might almost have been utter strangers. In fact, I've often wondered why we didn't give it up and leave the past buried. Because we were both a little obstinate, perhaps—I don't know.

"Now, in private life I was an auctioneer with a well-known London firm, and in the old days I'd often done

Adair small favours. For instance, if I knew he was interested in a particular lot I'd keep it back till he arrived if he
was late, and notify him about it beforehand. Well, he finally settled down in England in 1933, as I dare say you
know, and found the time hung pretty heavily on his hands.
As well as his place at Dorking he ran a small flat in
Knightsbridge, and when he got tired of the country he'd
pop up to town for a week or two, and as likely as not attend
the sales I was conducting.

"Now, I don't want to spin the thing out, Inspector. Adair
was interested in all sorts of antiques and curios, but particularly in enamel-ware. He's got—he had—a very fine
assortment of snuff-boxes, and so I knew he'd appreciate a
catalogue when Lord Howland's collection came up for sale
early in 1934. He duly turned up before the actual day, to
look round and choose what he'd bid for, and one item in
particular took his fancy. If he could get it for a hundred
pounds he'd have it, he told me, but not a penny more. If he
went on bidding after that it would be through sheer force
of habit, and I was to take no notice.

"That was another thing over which I'd often obliged him.
He'd set himself a limit, and when it was reached I'd stop
seeing his bids. He always made them the same way, by
winking his left eye, and so it was easy enough for me not
to look.

"Well, you can guess what happened. The article in
question was knocked down to somebody else for a hundred and ten pounds, and Adair was furious. He swore he'd
never told me to ignore him when the hundred was reached,
called me all the names he could think of, and then went
behind my back to the firm who employed me and began to
kick up a fuss there. Fortunately—as I thought—they didn't
take much notice: though the truth was that they intended
closing down in a couple of months, and didn't give a hoot
whether Adair had been done out of his snuff-box or not.

"However, he went on making such a commotion that in
self-defence I went to a solicitor. It's hard to say now
whether I'd really have gone through with a slander action
or whether I wouldn't. At all events, the upshot was that he
suddenly cooled down as quickly as he'd flared up, apologized, insisted on giving me a hand later when things began
to go wrong with me, and generally tried to make amends."

And thereupon Montague stopped, sighed, and regarded me intently from under his eyebrows.

"Does my story find credence?" he enquired presently in slightly acid tones, as I said nothing. "It's the truth, I assure you."

"Yes, I think I believe you," I told him, "It sounds all right, anyway. I presume your withdrawal of the action was the good turn you mentioned this morning—the one it was up to him to repay?"

"Yes—I was advised that I might have got pretty heavy damages out of him. As I've just said, he had repaid me in part, but I confess that I hoped I hadn't come down here for nothing. Anyway, it must be perfectly plain to you now that I had no motive after all."

"Now don't push conclusions down my throat, please," I said. "I haven't stopped asking questions yet. In fact, I'm going to repeat one I put this morning and if you give me the same answer I shall be annoyed. When were you last in the study, Mr. Montague?"

His face immediately became grave, and for quite two minutes he remained silent.

"All right, I'll tell you," he agreed with a nod and a rather defiant smile. "But I'd honestly forgotten that when I said just now that I'd only told one lie. It was on Thursday night, after the bridge. I'd just lost rather a lot of money, for me—over four pounds. Quite half of it was due to his appalling play, too—he had bad cards, and lost his temper over them, and that's fatal. I was a bit cross, especially as he wasn't a penny out of pocket over the evening—Tony'll explain if you don't know what I mean: so I tried to get some of my losses back from him. It was a stupid thing to do, of course, with him in the state he was, and I might just as well have saved my breath. I was only in the room for a bare two minutes, and how you guessed I was wrong in saying Tuesday I can't imagine. How?"

"You gave the fact away yourself," I said. "In that rather fantastic theory you outlined before lunch you several times mentioned the partition in front of the window, and in enough detail to show you must have seen it: yet the thing was only rigged up on Thursday, after the attack on Andrews."

He smiled again, in a somewhat forced manner.

"You're sharp," he informed me. "I'm glad I've nothing else to hide."

"I hope you haven't," I said, as I got up and stretched my legs. "I'll leave you alone for a bit now, but first there's something I ought to tell you: and something not without its possible reference to the subject of motive, either. This afternoon I found what purports to be Major Adair's last will, and if it is then you'll be the richer by five thousand pounds."

His expression when I left him was one of blank amazement. A moment later, purposely, I returned to ask if he knew anything about a diary. He said he didn't, speaking in a far-away voice as if his mind were wholly on my news concerning his legacy.

XXI

WHEN I got back to our room, Tony was still frowning over the directions for finding the treasure.

"This stuff's the sheerest drivel," he declared. "The man wants you to go round in small circles. Well,"—as he saw me smile, "you make sense of it."

"Maybe I will tomorrow," I said. "I've got half an idea of what old Jasper was driving at."

"Yes? I'll hold you to that. What happened with Montague?"

I told him, which took about quarter of an hour, and by that time it was midnight.

"And now I suppose you'll want to go to bed?" queried my friend, not very enthusiastically.

"No—I probably ought to be tired, but somehow I'm not. Let's do the old trick of making a list of people and motives and opportunities and all the rest of it."

"All right," he agreed readily: "only motives and opportunity for what? Inducing Adair to blow his head open? That's what happened, you know—after Lina's disclosures tonight it sticks out a mile."

"I know it does," I said: "but there can be times when a mile's too long. And not everything points the same way."

"No? What doesn't?"

"Buck's little discrepancy, for one thing: that hasn't been explained yet, remember, and if you say it was a mistake I shan't believe you. For another, how much noise do you suppose Hinkson would make in Montague's room?"

"Why, as little as possible, I should think—he was only two doors away from the man himself."

"Exactly—then why didn't he hear the shot? Montague's room comes in the lot from which he could, this afternoon."

"Oh lord, I don't know. Unless—wait a bit: yes, that works out. Assume he didn't hear it—if he did then he's still telling lies, and you can't go by anything he's said. Then surely we can time the murder—I mean the suicide—almost to a second. Hinkson began to go up to his own room to fetch his coat promptly at 9.45, and it probably took him about a minute and a half. During that time the other two were still in Lina's room, out of earshot same as him, and likewise Tilly, and Andrews and Cook had their wireless on. There you, are: everybody accounted for. At least, everybody except Buck, and I don't see that it matters much about him."

"Why not?"

"Well, he couldn't have done any murdering in the house because he wasn't in the house—he was out in the grounds."

"Not good enough," I said: "the position's not quite as simple as that, and nor am I. Seemingly nobody could have done a murder because of the bolts and padlock—therefore it wouldn't have been harder for Buck than for the rest of them."

"Oh, I see—that does clear things up. When a deed has two explanations, one possible and the other impossible, the first thing the true detective does is to discard the possible and obvious explanation. Then he suspects everybody within a mile and a quarter: logic by the yard, as you might say. If I asked you why you won't take things at their face value, I bet you couldn't give a coherent answer."

"I could," I averred. "I don't believe in the psychology of the suicide solution. Also, and equally relevant, there's the question of my peace of mind."

I then tried to explain to him what I have already set down earlier: about my bogy Cain, and my need to be entirely satisfied that for once the epithet 'sealed-room' wasn't synonymous with murder, as Montague had put it.

"Yes, I see your point there," he agreed. "I admit that sealed rooms are devilish shady, but that's only because you've been reading too many detective stories."

"Probably," I said: "but it might be worth working out why that should have made me suspicious. Do you remember what Dr. Fell said in *The Hollow Man,* parodying Kipling? 'There are nine and sixty ways to construct a murder maze, and every single one of them is right.' "

Tony shook his head.

"Sorry, I must be dull—I don't see the application."

"I mean that nowadays nobody dares write a sealed-room mystery unless he's found—or thinks he has—a new method for eventually unsealing his room. Or, alternatively, a new way of tricking the reader into believing that the sealing ever mattered a hang. Yet people do go on writing such mysteries, and the last few pages duly contain the requisite explanations; and extraordinarily ingenious some of them are.

"All right: now suppose that one day somebody works out still another fresh way of wrapping up the same old goods, and decides not to write about it but to put it into practice. Then where will you be? In the position of the reader who hasn't managed to make head or tail of the first two dozen chapters, and relies on the twenty-fifth to prevent his incipient headache. Only, this time, there won't be any explanation."

"Yes, that wouldn't be too pleasant," he admitted.

"Your *sine qua non* suddenly becomes a *non sequitur.* And yet there's never been a sealed-room murder in real life, has there?"

"Not one that I can recall off-hand. But that only means that for once truth lags behind fiction. The thing I want to be certain of is that this time it hasn't caught up and left me gaping."

"Very well—objection sustained," said Tony. "And now explain about not believing in the psychology of the suicide idea. You mean what Montague told you about Adair's being a coward?"

"No, or only partly: I'm relying more on my own opinions than on other people's. The trouble to begin with was lack of apparent motive for suicide: or the presence of abundant reasons why Adair should want to go on living, if you prefer it that way. Well, we've just been given a suicide motive, and my trouble now is that I don't believe there was time for it to work: I don't believe he had long enough to get into the right state of mind to shoot himself.

"Disregarding the question of why Hinkson failed to hear the shot if it was fired after he climbed into his tree, the absolutely latest limit for time of death is a minute or two before ten o'clock. Now, at 9.15 or thereabouts Adair threw Lina out, locked and bolted the study door, and began muttering. From the account we've had, the predominant factor in his emotional condition was fury, against Lina in particular and the rest of the household in general. Agree? Right: then without setting myself up as a psychological expert, I can't help wondering if many people commit suicide because they're in a rage with other people. Don't they do it rather when their rage has died down and self-pity has taken its place? When they've had time to convince themselves that they've nothing left to live for?

"What I'm keen to know is, can that change of outlook possibly have happened with Adair in about forty minutes at most, and probably in less? It's all a question of character, of course: some people's spirits rise and fall like a cork on a rough sea, and others stay in approximately the same mood for a week. What's your opinion about Adair? You knew the man to some extent: would there have been time?"

Tony shrugged thoughtfully.

"It's a sound point," he said, "but nobody in the world can tell you the answer. In my experience it didn't take much to make him cross, but a good deal to put him in a good humour again. He was a man of moods all right, but he wasn't exactly mercurial. If there weren't so much evidence to show he did commit suicide, I'd be inclined to say that he wouldn't have cooled down against Lina for some hours."

"And you agree that till he had cooled down he'd probably have been more likely to use his gun on her than on himself?"

"Yes, I think so. The trouble is that we don't know how much store he set on her honesty, of rather on his theory that she wouldn't reproduce her father's bad habits. Perhaps he'd grown to be absolutely certain he was right, and it came as a complete slap in the eye to find her a thief after all."

"And perhaps not," I said. "According to Lina's story he didn't 'find her a thief': he accused her of being one the moment she opened her mouth and said she'd lost the ruby. That seems to show he couldn't have been so very strongly convinced."

"Yes, that part's puzzling. Usually he'd purr at her even when he was snarling at the rest of us. Maybe he could forgive her anything but theft."

"In which case, why put temptation in her way by allowing her to keep the stone? He knew she'd got it—you told me he asked at dinner on Thursday."

"Yes, he did. The only sensible explanation I can think of is that the ruby was to be the great test of her honesty. If she guarded it well and truly, and returned it in due course, she and the theory would both receive full marks. If she didn't, out they'd both go."

"And himself as well? And wouldn't he set his test *before* he made a will in her favour? Not over a year later, surely."

"Now you're just being awkward," said Tony reprovingly. "And it'll all come back on your own head, not on mine. I think the man killed himself, and I'll lay you eight to one in cigars you don't prove different."

Before we went to bed we completed a summary of the situation as it appeared at the moment, together with various notes and remarks. For the purpose of this book I have somewhat elaborated our work, to make it more intelligible.

Events at Mauberley Grange on the evening of Friday January 21st, as described by different witnesses.

(V = place from which shots fired in study are audible
X= place from which shots fired in study are inaudible)

(1) Dinner finishes at approx. 8.50, whereupon
     (a) Hinkson returns to the study to receive instructions about tree-climbing and fetching Lina.
     (b) Tilly goes to her own room (X) and remains there.
     (c) Lina and Montague go to Lina's room (X), according to pre-arranged plan between her and Hinkson.
     (d) Andrews clears away.

(2) 9.00—9.05

> (a) Hinkson summons Lina, stays a minute or two chatting to Montague, then goes along to Montague's room (V) and begins to ransack it for the ruby. Apparently he did this skillfully, since M. says he noticed nothing unusual afterwards: but may not have realized import of question.
>
> (b) Lina goes to study.
>
> (c) Andrews and cook begin to listen to wireless in kitchen (V or X).

(3) 9.05—9.15

> (a) Hinkson searching for ruby (V).
>
> (b) Montague waiting for Lina in her room (X).
>
> (c) Lina in study being rowed by Adair, who accuses her of theft as soon as she says the ruby is missing. He then destroys the papers relating to his translation of the clue, but not the parchment itself, nor writing found in Macaulay.

(4) 9.15—9.17

> Lina's face is smacked, and she is ejected from study. From outside in the corridor she hears Adair lock and bolt the door. As far as she remembers, the desk was not particularly tidy, the partition was padlocked, and the decanter was on the mantelpiece half full. No pistol was visible or mentioned.

(5) 9.17—9.18

> Lina returns to her room (X), and remains there with Montague, as per her arrangement with Hinkson.

(6) 9.18—9.45

> (a) Hinkson in Montague's room (V) searching.
>
> (b) Adair (presumably) alone in study.
>
> (c) Lina and Montague 'chat' (X).
>
> (d) Andrews and cook listen to wireless in kitchen (V or X).
>
> (e) Tilly remains in her own room (X).

(7) 9.45

> Following arrangement with Lina, Hinkson leaves Montague's room (V), collects his coat from his own

(X), then goes downstairs and climbs tree at end of lawn near study window (V). *(N.B. It is suggested that Adair killed himself during the time that Hinkson was fetching his coat, everyone in the house then being out of earshot of the study. The probable duration of this period is 9.45½ to 9.47.)*

(8) 9.48

Lina gets rid of Montague by excuse about headache, and sits on her bed thinking (X).

(9) 9.48½

Montague back in his own room (V)

(10) 10.00

(a) Andrews leaves kitchen—where wireless is turned off at time-signal—and proceeds to study with hot milk and sandwiches; but can get no answer.

(b) Tony and Beale enter house and go upstairs, just missing Andrews in hall.

(11) 10.11

Beale is called out by Andrews on false pretext.

(12) 10.12—10.19

Explanations, attempts to gain admission to study by knocking, and inspection of window from outside.

(13) 10.19—10.20[1]

Door broken down; discovery of body; Hinkson fires three times from outside at unknown intruder, then scrambles down from tree and gives chase.

(14) 10.20—10.22

During a minute and forty seconds of this period Buck says he ran from woods to house; but his story may be taken as untrue as far as the time is concerned.

(15) 10.38 approx.

Hinkson arrives indoors, dirty and breathless, having seen no one but Buck.

---

[1] Details Chapter XII

(16) 11.00 approx.

Buck arrives indoors, having seen no one but Hinkson.

Addenda

(17) During the period 9.00 to 10.20 Buck was alone in the grounds carrying out his usual job, according to his own statement.

(18) The Judds say that they remained indoors at the lodge from 7.00 onwards, but their manner is not particularly convincing.

Remarks about Motive.

Nobody has expressed any marked sorrow at Adair's death, but this is not perhaps to be much wondered at. If it could be shown to be murder, there would be seven natural suspects: Lina, Tilly, Hinkson, Montague, Buck, Andrews, and Judd. (Mrs. Judd and the cook seem not worth serious consideration.)

No known motive can be assigned to Buck, Andrews, or Judd; in fact, each apparently loses rather than gains by his employer's death. None of them receives a penny in his will. With regard to the other four:

(a) Lina benefits financially to the extent of a great many thousands of pounds, and on her own admission she was so desperately hard up that she stole the ruby entrusted to her care.

(b) Montague benefits to the extent of £5000: and he too was in need of money. He may also have borne Adair a grudge.

(c) Tilly in actual fact gains £1000 by her father's death, but she may well have hoped to gain far more. In addition, she is suspected of having loathed him—and not without reason.

(d) Hinkson receives nothing under the will, but he was apparently the only person at all in Adair's confidence about the clue. Is it possible that he had interpreted it, knew the location of the treasure, and removed Adair as the essential preliminary step to obtaining possession?

Further suspects to be considered:

(e) The unknown person who attacked Andrews on Wednesday evening.

(f) The unknown person at whom Hinkson fired three shots.

(g) Warner, Adair's previous secretary, who may be *(e)* or *(f)*, or the accomplice of either or both. In the latter case, account must be taken of Buck's suggestion about a possible spy in the household.

It seems safe to say that the likeliest general motive would be one connected with the treasure.

Remarks about Opportunity.

Only Andrews, of the seven potential suspects, has a witness to his movements during the whole period 9.00 to 10,00. With regard to the others, corroboration of their movements is lacking in respect of

(a) Buck for whole period;

(b) Tilly for whole period;

(c) Judd for whole period—disregarding his wife's evidence;

(d) Hinkson from approx. 9.05 onwards (after he left Montague in Lina's room);

(e) Lina for periods 9.05-9.18 and 9.48-10.00;

(f) Montague for periods 9.05-9.18 and 9.48-10.00;

Remarks about Conduct.

Lina, Hinkson, and Montague all told a number of lies to begin with, but have since given reasonably satisfactory explanations for so doing, the first two voluntarily. Buck also told one patent lie, but has not offered, nor been asked for, an explanation. It is possible to imagine that Montague wished to conceal knowledge of the range at which Adair was shot.

General Remarks.

If Adair was murdered, he must have been drugged first: he would never have let another person put a loaded pistol into his mouth. Montague suggested the idea: can he conceivably be so cock-sure of his safety that he thinks he can afford to give away important details?

It was not considered politic to ask him why he lied about when he was last in the study. The obvious reason—fear that if he were known to have tried to borrow money unsuccessfully he would seem to have a motive—appears to make his lie the action of a nervous innocent man rather than of a guilty one.

Attempt to deal with the question—Can Adair's death possibly have been murder?

Facts which may be relied upon:
(1) Adair met his death, in the study.
(2) He was shot by a .25 automatic pistol, the muzzle of which was inserted in his mouth. It is thus clear that a silencer could not have been used.
(3) Death was instantaneous: Adair could not have moved again, after the shot, of his own volition.
(4) The study contains only two entrances for human beings, the window and the door.
(5) The window could not have been used for a murderer's escape, being shut off by a partition which was padlocked on the inside. This padlock can be fastened only with a key, and the only known key was in Adair's pocket. While this does not rule out the possibility that there are or were others in existence, nobody in the world could have closed the padlock from the window side of the partition. Furthermore, the windows have obviously not been opened recently.
(6) The door could not have been used for a murderers escape, since it was both locked and bolted from the inside. The bolts are too stiff for there to have been any trickery from the corridor.
(7) There is not and was not any hiding-place in the room that would stand a moment's scrutiny.
(8) The paper found in the Macaulay was written by Adair.

Facts which are doubtful:
(9) That all the witnesses have now been wholly truthful.
(10) That the paper in the Macaulay was put there by Adair.
(11) That it is a true translation of the last two paragraphs of the parchment.

(12) That the bullet found in Adair's head came from
      the .25 Webley automatic found on the floor near his
      chair: but it is a million to one it did.

*The only reasonable conclusion seems to be that Adair's
death was suicide.*

"Unless," suggested Tony "anyone could have sneaked
out of the room while you were staring at the body."

"Impossible," I said. "The only place for anyone to have
remained out of sight even for a moment was behind the
door, and Andrews had the wit to look there as soon as he
followed me in."

"How d'you know?"

"I saw him: and nothing thicker than a shadow could
have edged between us while I stood staring, as you put it."

"All right: then will you once and for all stop talking
nonsense about murder?"

"Yes, I think I'd better: unless my funny business with
the book-case comes off, in which case I'll be pretty sure it
was murder."

"Yes: but it won't come off," declared Tony. "If it does I'll
pack my bag and go while I'm sane. You see the only
possible murderer left, don't you?"

"No" I said, falling into his trap. "Who?"

"Why, the ghost of Jasper Mauberley, of course."

XXII

THE first thing I did after breakfast the next morning,
Sunday, was to warn Johnson that he must keep out of the
study. I didn't want anybody who might feel so inclined to
be put off scrutinizing the Macaulay by his presence. The
second thing I did was to go outside the moment I caught
sight of Judd from the windows, and approach him with a
pleasant smile.

"Good morning, Judd," I said. "Is your wife at home?"

The question was meant to disturb him, and plainly did
so. He regarded me with an expression that was half glare
and half frown, spat viciously for no apparent reason, and
assumed the attack.

"And what's that got to do with you?" he demanded, in tones far removed from yesterday's amiability.

"Now don't go taking things the wrong way," I said. "I've no improper designs on the good lady—I merely want to be sure that I'll find her in if I go down to the lodge."

"What for?" he queried immediately, suspicion in his voice and manner, and something in the nature of a threat too, I thought.

"Strictly speaking, that's my business," I returned, "but I don't mind telling you. I want to find out what you were really doing on Friday night, and I think perhaps I can induce her to say. Unless, of course, you'd rather explain for yourself?" I added, after a suitable pause.

For a moment I thought he would either bid me go to hell, or put his head down and charge; but my apparent confidence decided the issue. After a little skirmishing: he agreed to talk, and I led him beyond earshot of the house. I'm sorry if anyone has been expecting fireworks: for myself, I felt very well satisfied, because I was nowhere near his weight.

"Incidentally, I shall see your wife anyway," I warned him, when we were undeniably alone, "so whatever you say had better be the truth. What did you really do on Friday evening from nine o'clock onwards?"

His story was very simple, and confirmed Tony's suspicions. For some days past he had been encouraged by Lina into considerably greater familiarity with her than Adair would have approved of, and on the evening his employer met his death he had gone up towards the Grange about half past nine in the hope of a few minutes in her company. From the orchard, however, he had seen that she had Montague in her room with her, and so after hanging about for a short time he had returned to the lodge. He had been all the while in front of the house, not at the back, and could tell me nothing about Hinkson's tree-climbing activities. I pressed him to be more exact about times, and he finally decided that he had probably set out at 9.25, stood watching Montague and Lina from 9.30 till 9.35, and got home again by quarter to ten. He had seen no one moving in the woods, and had heard no shots fired but accurately timed the arrival of Tony and myself in the Bentley as having occurred at five minutes to the hour.

He seemed not to mind revealing his own duplicity towards his wife, nor Lina's towards Hinkson: or perhaps her triplicity, in view of his assurance that Montague had been kissing her 'good and hard'. He was certainly aware that there was something between the girl and the secretary, and declared now that should have nothing more to do with her.

"A proper little bag o' trouble she is," he said, spitting again. "Gawd, I wouldn't 'alf lay it across 'er if she was my gal."

As soon as I had finished with Judd I went down the drive to interview his wife, and obtained from her a confirmation of his new story as far as times went. It also appeared that Mrs. Judd was pretty sure what her husband had been up to with Lina; but today there was no sign of the resentment which Tony had witnessed earlier in the week. She just sat gloomily licking her thin lips, a haggard unloved woman with neither looks nor hope left.

On the whole I was inclined to believe what I had just been told, I thought, as I made my way back to the house. All the same, I still wasn't quite certain of Judd: but that might be because he was so personally distasteful to me.

And then, a moment or two later, I saw something which interested me profoundly: Lina and Buck together again, walking through the trees. I say I saw them, but I think it was the sound of their voices which first attracted my attention. There was no question here of friendly conversation: it seemed to me from where I stood watching that they were engaged in some sort of disagreement. I wasn't near enough to hear actual words, though, and debated for a few seconds what would be best for me to do. If I attempted to approach I should probably make them aware of me, and anyway I had little reason for supposing that I should overhear anything worth while; yet there was always the chance of doing so, and I can't say I had any scruples about eavesdropping on either of them.

My difficulties were soon settled for me, however. As I began to walk on, pretending not to notice them, I saw Buck look up and spot me, and nudge Lina. All this out of the familiar corner of the eye, be it said: I have at one time and another taken some pains to practise the art of seeing without seeming to. There followed a stifled exclamation and a scuffle, and when I glanced round (as I had to do to keep up my pretence) Buck was alone.

"Hullo!" he said, without advancing towards me.

"Hullo!" I answered. "Have there been any telephone calls for me, do you know?"

"Couldn't say at all—don't think so." And with that we went our separate ways.

I was in fact expecting a call from Relf, and it came about eleven o'clock. He had been unable to trace the presence of any suspicious strangers in the district, and the report on the decanter and glass was entirely negative. The only fingerprints found were those of Adair himself, as also on the pistol. The ashes from the grate were still being examined, and the bullet, so that I could get no help there yet, but the police-surgeon was with him, in case I wanted to ask any questions.

I had five minutes' talk with the man, who had completed his autopsy but found nothing worth a comment.

"No trace of drugging?" I queried. "Did you think to look? I meant to suggest it."

"Yes, as a matter of fact I did think, but there wasn't anything."

"Which means that he definitely wasn't murdered, I suppose."

"Why so?"

"Well, you wouldn't let anybody put a loaded pistol into your mouth if you were awake and sane, would you?"

"Not often: but because I found nothing, it doesn't follow that he mayn't have been doped, you know."

He then proceeded to give me some potentially valuable information. By the use of one of the barbituric hypnotics, he said, such as *Nembutal,* administered orally, Adair might have been put into a deep sleep which would last from six to eight hours. Also, provided the administration took place long enough before death, say five hours or more, the drug would be absorbed into the blood-stream sufficiently to defy detection afterwards.

"Administered in what?" I asked.

"Oh, food—preferably something with a bit of taste, like curry or bad eggs."

"Would whisky do?"

"Well, hardly—or only if you gulped it down."

"And how much would you give?"

"Oh, the normal B.P. dose is from one and a half to three grains, but I think I'd be on the safe side and give five. And I think I'd choose one with bromine in it, to avoid any possible preliminary excitement: *Pernocton,* for instance."

"And could an unauthorized person get hold of the stuff?"

"Ah, that's another matter: but people do get hold of things they aren't supposed to, you know. I shouldn't let that part worry you. Dozens of doctors have had things pinched from their cars or surgeries."

"Including yourself?"

"No—I never lost more than a bottle of dill-water."

About quarter of an hour later I received a second telephone call, from New Scotland Yard. It was Superintendent Vinney speaking the other end, the man under whose immediate supervision I usually work, and he had an item of news for me which was well worth hearing.

"You were asking about a fellow called Warner, weren't you?" he said. "Adair's previous secretary."

"Yes sir—have you found anything?"

Vinney chuckled hugely, making my left ear buzz.

"I'll leave you to decide that," he told me. "Warner got twelve months' hard labour last August for stealing a small quantity of jewellery and a cheque-book from a clergyman he'd duped. In Wolverhampton, that was."

And now, about midday on Sunday, after I had been in the house for something over thirty-six hours, I reckoned that it was about time I began to keep my mouth a little more closed and my eyes and ears a little more open. Consequently I spent the afternoon dozing in the intervals of profound contemplation upon the subject of Adair's death. The reason for my inactivity was that I wished to spread abroad the impression that I wasn't unduly worrying myself about matters. The result of my thinking was that I could see only one thing which would save me from abandoning all consideration of murder: clear proof that the paper between the pages of the Macaulay had been tampered with.

Nevertheless, there were still several points that I didn't pretend to understand: in particular, the reason for Buck's lie, the attack on Andrews, and the identity of the mysterious stranger at whom Hinkson had fired. As I believe I have said, there were traces of someone's having moved about by the rhododendron bushes, though nothing helpful in the way of a footprint or other clue.

The next two days I'm not going to describe in detail. I sat about, chatted amicably with anyone who seemed inclined for a chat, and half a dozen times a day surreptitiously inspected the book-case in the study, without result. The paper had not been moved—of that I was sure: and I began to think that nobody but Tony and myself knew of its existence. Without arousing possible suspicion it was impracticable to keep an eye continually on the room, and so I was compelled to leave it unattended. If my trap ever worked I should have to rely on my wits to pull me through.

As far as I could manage it I made a point of seeing something of everyone during those two days: and succeeded except in the case of Hinkson, who was really devoting a lot of time to the parchment. When I asked how he was getting on he confessed that the task was proving harder than he had anticipated. The three or four lines he showed me tallied well with what Adair had written, but naturally I kept quiet about this.

As will perhaps have been noticed from my lack of reference to her, Tilly had kept out of my way since our talk on Saturday. On Monday, however, she sought me out of her own accord, and put a somewhat curious point of view before me.

"Inspector," she said, very earnest and unattractive with her eyes looming enormously behind her thick glasses, her hair a mess, and two inches of frayed petticoat showing, "I've thought of something that may be of importance. On Friday morning my father happened to mention in my hearing that you were bringing down a copy of Kipling's *Barrack-Room Ballads*. He had had a phrase from one of them on the tip of his tongue for days, and was worried because he couldn't verify the quotation. Now, please don't make fun of me for what I'm going to say, but he simply wasn't the kind of man to shoot himself until he *had* verified it."

I attempted to preserve a serious demeanour, and listened attentively while she elaborated her psychologically absurd theory, promising that I would think about the matter, though mildly suggesting that perhaps her father had managed to remember the quotation for himself. I then took her completely off her guard with a snap question: perhaps unwisely, but for once I spoke on the spur of the moment.

"You hate Lina like hell, don't you?" I asked.

For a moment she turned fiery red, giving a pronounced gasp. Then she set her thick lips tightly and nodded two or three times.

"I absolutely detest her," she said at last in a low trembling voice, her hands gripping the arms of her chair. "She's a dirty immoral little liar."

"A liar about what?" I pursued quickly.

"Saying I took the ruby," came the answer, with equal speed and a good deal more heat. "I didn't. I tell you I didn't!"

"Hush!" I said, for she showed signs of becoming hysterical. "*I* know you didn't. Don't let it worry you any more. Just answer me one other question, and then go and lie down for a bit. Did you tell your father that Lina had practically accused you of theft?"

"Oh no!" she exclaimed. "Of course I didn't—I wouldn't dream of sneaking like that."

"Thank you, Miss Adair,"—with a smile. "That's really all—and keep your end up."

Montague passed most of the time he was downstairs in writing letters, or reading, or nodding over his book. He stayed in his room until lunch, and returned to it again after dinner, and on the whole didn't prove a very sociable companion. He gave me the impression that he was still secretly annoyed about the exposure of his visit to Lina's bedroom. When I asked if he'd done any more towards building up a satisfactory murder theory he smiled and shook his head.

"You pulled the other one to bits, so I think I'll stick to the *Newgate Calendar*," he told me.

Andrews too I managed to talk to: the man interested me. I was scarcely more successful than Tony had been at getting him to be communicative, though. He remained deferential and reserved, and underneath everything still a little frightened, I thought, as if he had not yet fully got over the noise of Hinkson's revolver shots. I do know that an unexpected firearm report can profoundly disturb people who are gun-shy, and so found his manner excusable, in the circumstances. As far as I could tell, there wasn't a single member of the household, bar the cook, with whom he ever entered into conversation on his own initiative.

Judd had resumed his somewhat care-free attitude to life, and seemed not to avoid me unduly: certainly not so pointedly as he avoided work, spending the majority of his time either at the lodge at leisure or loafing about the grounds with a cigarette dangling from his brutal mouth. He did offer to clean Tony's car, and made a fair job of it: but at his own pace, polishing the lamps as deliberately as if they had been fragile precious stones.

The two persons whose behaviour struck me as most worthy of note were Buck and Lina. The girl seemed to have changed greatly in the last two days, and Tony bore out my opinion. For one thing, she no longer troubled so much about her personal appearance: she was still neat and well groomed, but not flauntingly *chic.* As well, she was most definitely unsociable. This might have been because Hinkson was so busy with the parchment, of course, but I wasn't sure. By all reports his previous seclusion in the study hadn't affected her like that. To Montague she was almost deliberately rude, seeming to snub his slightest approach to familiarity. Tony and myself she ignored, or smiled at mechanically and meaninglessly. With Tilly I never saw her alone, Judd and Andrews she looked through as though they had been walking windows, and only to-wards Buck did she condescend to be anything like amiable.

The little crook, with his bald head and bad teeth, wasn't by any means a physically attractive man, I should have thought, even to a person of such varied tastes as Lina. Common sense suggested that all the active part of any friendship between them would derive from him: yet I was still doubtful. Two or three times I saw her stroll up to him uninvited, if with no great show of pleasure, and it occurred to me that possibly Buck had found some remunerative way to worry her about the episode with Tilly and the stream, or even about her association with Judd. I wouldn't have put blackmail past him for a moment: I felt sure he was a rogue, although an engaging one. I took the opportunity on Tuesday to examine his gun: it was a .38 revolver of American make with a bone handle.

In short, there were very few developments of any kind at Mauberley Grange between Sunday midday and

Wednesday morning; or, if any took place, they did so unobserved by me. Of extraneous ones it is sufficient to mention the following. Through Adair's bank I got in touch with his solicitors, and learnt that his estate would amount to at least a hundred and thirty thousand pounds. Through Scotland Yard I communicated with Hyams, the money-lender to whom Lina said she had sent the ruby. He confirmed her story, adding details: she had formerly owed him £290 odd, and he had received the stone on the previous Thursday morning in an envelope registered from Gloucester on the day of her visit there with Tony. I also heard from George Relf that the bullet extracted from Adair's skull had undoubtedly been fired from his own Webley, and that there was nothing suspicious about the ashes from the grate.

A part of those two days, I may say, I devoted to the supposed translation of the clue to the treasure, and soon satisfied myself that it was genuine. As I had told Tony, I wasn't without ideas about how to wrest a meaning from its apparent foolishness, and these I proceeded to test. It took me, I judge, about four and a quarter hours to determine—theoretically—where Jasper Mauberley's legacy to future cryptographers had been placed: though that didn't mean it was still there, of course.

If I refrain till later from explaining how I arrived at my conclusions, and what they were, I do so in case anyone wishes to attempt the solution of the clue for himself. To make the task as easy as I can without making it wholly simple, I will say that the necessary equipment consists of two books intimately connected, and a not too disciplined imagination. Nor is there any need to take over four hours: everything could be done comfortably in twenty minutes—with the right kind of imagination. Bear in mind that Jasper Mauberley was a shrewd far-seeing man, secretive, educated, and not unsubtle.

XXIII

It was at ten o'clock on Wednesday morning that I made the discovery which transformed the whole aspect of the case. Without much hope I went into the study to look at the paper in the Macaulay, and so familiar had I become with the exact position of the books on that particular shelf that I saw at my first glance that the spacing was different. Experiment convinced me that there was no possible doubt: the Macaulay had to be actually opened before the margin of five millimetres between paper and book-edge could alter, and it was now a good eight millimetres at the widest part. After so long a wait the apparent working of my trap worried me. I had a short discussion with Tony, who seemed impressed, especially by the absence of finger prints on the shelf. As far as we could determine there should have been some, especially if the book were—almost inconceivably—displaced by a casual hand. Nevertheless he still maintained his contention that murder was impossible.[1]

"How *could* it have been?" he asked: the vital question become now suddenly so insistent.

"By a trick," I said. "All conjuring is a trick."

"Yes: cursed are the sleight of hand, for they shall escape arrest. And the sleight of mind too."

He nodded at his own wisdom and turned to the door, stooping down to measure the gap under it.

"The only solution we haven't seriously considered is murder at a distance," he told me. "The room was genuinely locked and bolted from inside by Adair himself; and then, miraculously, he was killed from outside—by a trick, of course. D'you know what I saw in Tilly's room yesterday?"

"*The Magician's Mystery Box?*" I suggested.

"No—a fishing-rod. And the queer thing is that she didn't know it was there. It belonged to Adair, and she thought it was in his room, but she happened to go to her wardrobe, and found it in the corner: a sectional affair about fifteen feet long when it's fitted up, very slender and whippy. The reel was eighteen inches from the end, and the rest of it would go under that door all right. And the desk is eleven feet away—I measured it this morning early."

"All right," I said, "run along and work it all out, and then come back and show me how it was all done. And why a fishing-rod? Why not a butterfly-net, or a long intelligent snake? Yes, I know I said trick just how, but I really meant illusion. Please yourself, though—but do go away."

Well, I sat, and I thought, for some two hours. This clearly isn't the place for a detailed account of my cogitations, however, so I will only say that by lunch-time I had managed to work out a possible answer to the question of how Adair could have been murdered. I was helped by two things: the discovery by reflection of yet another lie still outstanding in the evidence I have recorded—and one which I blamed myself for not noticing at the time; and realization that I might have been very cleverly duped into helping the author of the crime unwittingly.

The more I considered my new idea the more I liked it, and after lunch I began to ask questions again. From Tony I required a repetition of what everyone had done on Thursday evening, but learnt nothing apart from what he has related in Chapter X. It may be assumed, incidentally, that I was familiar with all the important details in his part of this book.

Next I looked for Montague, whom I found studiously playing chess in the lounge with Tilly, and apparently getting the worst of the game. When I got him alone I asked if he was still certain that Adair had been a coward.

"Oh yes," he replied. "Why shouldn't I be?"

"If he was expecting a burglar, and you offered him a bullet-proof waistcoat, he'd accept?"

"Certainly—and probably grumble because you hadn't two for him."

Tilly was my third informant, and she disappointed one.

"What was there for lunch on Friday, Miss Adair?" I enquired.

"For lunch?" she repeated. "But why ever do you want to know?"

"That doesn't matter: what was it?"

"Oh, sorry," she muttered, quick to take offence. "Roast mutton—mutton and potatoes and cabbage, and red-currant jelly, and rice pudding, and biscuits and cheese."

"And what to drink?"

"Beer and cider."

"Which did your father have?"

"I really don't remember: beer, I should think. He was in the dining-room for only a very few minutes, though, and then he and Mr. Hinkson went back to the study."

"I see—thank you. Now, Miss Adair, you told me the other day that you keep a diary, and I think perhaps you may be useful, if you're willing. Can you give me a fairly exact account of what happened about that burglary—when Warner's supposed to have photographed the parchment?"

"Why yes, I think so," she answered. "I'll run up and get my last year's diary, shall I?"

The following is a summary of the information she was able to give me, after consulting several pages of her thick solid handwriting.

On Friday, September 15th, 1937, Adair had brought them all down to Mauberley Grange by car to look the place over: herself, Lina, and Hinkson. They came direct from Dorking, and put up at a hotel in Gloucester. The week-end was spent in examining the Grange, and on Monday they approached a builder about the probable work required to make the house habitable. Then, on Tuesday, Tilly and her father went up to London for the day, also by car, and brought back with them two experts in seventeenth-century architecture with whom Adair was acquainted. It was during their absence, they learnt afterwards, that a telegram arrived from Andrews at Dorking saying that he suspected there had been a burglary there the previous night. On Wednesday the 20th, in consequence of this telegram, they all returned home, and later informed the police, but with no satisfactory result. As far as Tilly knew, between September 15th and 20th Adair himself had been the last of the four to leave the Dorking study, the room which contained the shorthand diary of Thomas Boon, and the first to re-enter it.

Andrews confirmed her part of his story, and explained that he had suspected a burglary because he had found a ground-floor window open which he was sure he had fastened himself before going to bed. Again, I must admit, I was disappointed: until I saw that I was only making difficulties for myself.

Then, leaving it as late as I dared, I took the logical next step. After fixing up appropriate arrangements by telephone with George Relf, at ten o'clock on Wednesday evening I arrested Buck on behalf of the New York Police, and watched him safely conveyed away by car. He struggled a bit, and swore considerably, and his look at me flashed with hostility, but he got no chance to reach for his gun.

INTERLUDE

And now, if only as a formality, I will put these questions to the reader.

Who fired the shot which killed Major Adair? How was his murder contrived?

The first isn't in my opinion particularly difficult to answer, and I am afraid that some people may think it a great deal too easy. The second may provide material for a few minutes' thought, though; but, in case not, here is a third.

What did I do about things next? From whom did I seek corroboration of my theories, and what means did I use to obtain it?

The following chapter contains my answers to those questions, and in an appendix I will include the full solution of the clue to Jasper Mauberley's treasure.

XXIV

WELL, the next thing I did was, considerably after midnight, to rouse Johnson, bid him dress, and then go along with him to Andrews' room. We entered, locked the door, and awoke the sleeping butler. He sat up in bed, a startled elderly man in flannel pyjamas, and he still wore his eye-shade.

"You'd better not make a noise," I said, putting my lamp down on the mantelpiece while Johnson took up his position against the wall. "We're going to have a long talk, Andrews—or rather, I'm going to. No,"—as he began to express stupefaction, "don't try that game: it's mere waste of breath and energy. All you've got to do is to lie still and listen."

Even in the dim light I could see that he was badly frightened, and I felt glad. I was playing a tricky hand, and should need all the help I could get. This is the story I told him, as nearly as possible in the words I used then; but where necessary I have amplified it a little so that all may be clear.

"I'm going to explain to you exactly how Major Adair was murdered, and by whom," I said: "though not because I think you need the information, I may add. But quite possibly I shall mention details that even you don't know, so you shouldn't be too bored. You can smoke if you like—my cigarettes."

I had purposely borrowed a packet from Tony, and now handed it to him: I was giving him no opportunity to poison himself, if he had ever thought of doing so. He thanked me gravely, seeming to have recovered a little of his composure; but his forehead glistened with sweat, and his hand trembled as he took the lighted match from mine.

"Please understand that at the moment I'm expressing my personal opinions," I went on: "and please interrupt only to correct me if I should go wrong.

"The author of the whole plot, then, was Hinkson, and his accomplices were you and Lina. Exactly what hold he has over you I don't know, but I imagine there must be one, or you wouldn't be in your present position. He was engaged by Major Adair to take Warner's place at the end of last May, and very soon decided that there might be something in this treasure of which Boon's diary spoke. Now, he's a man with a very fertile mind, of the criminal variety, and as far back as last September he set about laying preliminary plans for a hypothetical murder, or robbery at least. By that time he'd got you in his grip: over something to do with forgery, I shouldn't be surprised, from what Mr. Purdon saw once. He'd also been getting very friendly with Lina, and discovered that she was a person of his own kidney: that is, entirely without scruples. She'd go to any lengths in pursuit of her own interests, not stopping short at murder.

"His first step was the supposed burglary of the Boon diary: you faked the evidence while the rest of them were down here, and he suggested the photography business when he got back. Adair didn't take much notice, though, so later in the month the mysterious Mr. Duffy was intro-duced—the gentleman who telephoned the agents that he was willing to buy this place for £7000 hard cash.

"The purpose of all that September planning was to provide a suspect for any future crimes that might take place. The fact that Warner was actually in prison from August onwards probably wasn't known to Hinkson, but in any event he was covered. If Warner turned up with a cast-iron alibi, Hinkson would merely smile slyly and say that of course nobody expected the man to have done anything himself: he'd got accomplices to work for him.

"The next step took place at almost the same time, and seems to show definitely that the idea of eventually mur-dering Adair was present even then in Hinkson's mind. On September 18th, Tilly tells me, he and Lina were left behind in Gloucester while she and her father went up to town for the day. Hinkson seized his opportunity, even if he didn't engineer it. The two of them, I'm certain, spent the 18th here, finding out from which other rooms shots fired in the study were audible. Since they couldn't then know which would be the study, they doubtless made experiments with all the likely ones: Hinkson strikes me as the sort of man to do the job thoroughly, and it was the only chance he could possibly have had.

"At that time, of course, so far ahead, he had no detailed schemes, but they began to take shape when you were all settled in here. By then he knew how many people he'd got to cope with in addition to yourself and Lina. There were actually five, leaving out the Judds, who lived at the lodge: the cook, Tilly, Mr. Montague, Mr. Purdon, and Buck.

"Naturally he made no move until the clue to the treasure had been found, and naturally the ruby came as a surprise. It was one he turned to good account, though, and it shows the calibre of his mental equipment that he was able to do this so speedily. However: the moment the parchment came to hand he set about plotting in earnest: and really brilliantly, I think. Adair was to be murdered as soon as he'd translated it, and since Hinkson was working on the job he could gauge pretty accurately when that would be, once a start was made. By Monday night, I dare say, he knew it would be safe to kill the Major on Friday. If the man showed signs of finishing his clue earlier, Hinkson would mislay it, or do something to hang out time. If the thing wasn't ready by Friday, he would rely on his own ability to complete it for himself.

"Very well, then: Adair was to be murdered on Friday. How did Hinkson propose to do this? As I see it, he first made use of the ruby, in Lina's possession. She was instructed to post it off to the moneylender to whom she was in debt, and then—apparently—tell him she'd lost it, and ask his advice and help. That would give him an excuse for going to Mr. Purdon, in the hope of getting myself down here. He'd heard about me, of course, and it would materially assist his murder-plot to have the body actually discovered by an absolutely reputable person. If I'd refused to come, or been unable to, I don't doubt that Mr. Purdon would have been the dupe: as it was, things worked out just right for Hinkson—or wrong, according to the way you look at it.

"An important point to note is this. Hinkson could tell almost to a minute when the Bentley would get back here on Friday night, or at any rate the earliest possible moment, and arranged his murder to fit in with that time. There was no luck about it—just sheer cold-blooded deliberacy. If I'd come by an earlier train, then Adair would have died earlier; if later, later. Again, in spite of his subsequent account of his movements for Thursday evening, I don't doubt that he really did borrow the cook's wireless set, as he said he was going to do when he was asked to play bridge. His reason would be to listen to the weather-forecast, weather conditions being a possibly important factor in his plans. Snow or heavy rain would upset them; but there was neither.

"And now for how the whole appalling business was contrived—in my opinion, of course.

"After lunch on Friday Adair and Hinkson returned to the study, and as soon following that as possible Adair was doped. It took me some time to work out how that could have been done, when I learnt that there was nothing like curry for the meal, but the slowness was my own fault—I should have remembered the percolator and the tin of coffee earlier. That was the means employed, without a doubt, and it would be simple enough for Hinkson to per- suade his employer to make some, or let his secretary solicitously do it for him. 'You look tired, sir—wouldn't you like me to fix you up with some of your special Costa Rica?'

"Well, very soon after two o'clock Adair was dead to the world, and due to stay like that for some hours, roughly according to the dose given. Hinkson remained in the study continuously, of course, to keep out intruders, and the partition would effectively shut out the view from the garden.

"That partition, by the way, was probably Hinkson's idea too. Mr. Montague tells me that Major Adair was a coward in spite of his D.S.O., and would willingly listen to suggestions for his safety. As a reason for putting the thing up Hinkson made use of another essential move in his game—your black eye. You two arranged that between you on Wednesday night, and well enough for nobody to have suspected you. And the reason for your black eye? So that you'd have an excuse for wearing an eye-shade, thus making yourself easily recognizable—*especially to a stranger.*

"Going back to Friday: while he was having dinner Hinkson would lock the study. Afterwards Lina took Mr. Montague up to her room, which is one of those from which shots in the study are inaudible. Montague's own isn't, though, and that explains the necessity for the move. To account for it they evolved a complicated story, which they duly told me on Saturday night. It wasn't a bad story either, on the surface, though it had one or two weak points.

"That was two of the household disposed of—given an excuse for hearing nothing. You settled the cook and yourself by having the wireless on loudly, Buck was out in the woods, Mr. Purdon was in Gloucester, and only Tilly remained. Fortunately her room too was out of earshot, and as well there was no danger that she'd make a nuisance of herself. She got on notoriously badly with her father, and wasn't anywhere near friendly enough with Mr. Montague or Lina to seek them out for company.

"Now I'll come to the actual killing of Adair. It was meant to look like an unquestionable suicide in a sealed room: the man would appear to have shot himself with his own pistol, the end actually in his mouth, and the door and partition and window would all be above suspicion. Well, the shot was fired all right, probably about quarter to ten: but not by Adair. It was fired by Hinkson: *who afterwards remained in the study.* That fact constituted the master-stroke of the whole plot, and I freely admit it was one of the boldest stratagems I've ever met with.

"About ten o'clock Mr. Purdon and I returned from Gloucester as expected, and shortly afterwards you came on the scene, calling me out and spinning your tale about not being able to get an answer when you went along with your tray. You did your part very well, incidentally—your mistake in calling me Inspector Bell instead of Beale was most natural. You also had an ingenious reason for persuading me not to call Mr. Purdon: Adair was probably drunk, and you knew he wouldn't want to be seen like that by anyone he knew, especially anyone with whom he wasn't on particularly good terms at the moment.

"Now, I've already said that Hinkson remained in the study with the corpse; yet, for a reason which will soon become apparent, there had to be somebody in one of the beech trees—the one from which it was possible to see into the room through the chink at the top of the partition. That somebody was Lina, free after she'd dismissed Montague about ten minutes to ten. What made that part clear to me was luck, more or less. Mr. Purdon happened to see her buying some cheap black underwear in Gloucester, and you don't need to look twice at Lina to see that she wouldn't normally wear black bloomers. The colour and style of garment were chosen so that she'd be less visible from the ground in case there should be anyone about. Doubtless it was still an awkward moment for her—though not from reasons of modesty—when I walked round outside with you, but everything passed off all right.

"Indoors again, I began to break into the study after getting no answer from repeated knockings, and then came the climax of the plot. Let me try to recapitulate what happened. At my third charge the door went crashing open to my right, effectually shutting off all view of Hinkson crouching behind it. In I blundered, and my attention was at once arrested by the sight of Adair, in front of me and to my left, sitting dead at his desk. I stood still staring, as well I might, and though I might equally probably have moved forward, the one psychological certainty was that for a second or two I wouldn't move backwards or look behind me. However, to make entirely sure of this was Lina in the tree outside, her eyes—perhaps with the help of opera glasses—glued on the study window. As I proved for myself on Saturday night, it's possible to see the lower half of the door from one position, and that was all she wanted to see. As soon after my entry as she could manage it she fired Hinkson's revolver, thus assuring that my attention would remain centred in front of me; and, at second intervals, she fired twice more with the same purpose.

"She used a revolver, and blank cartridges, of course, so that there should be nothing in the way of bullet-marks or ejected cases to show that she'd been up the tree to the left of the window, and not in one of the pair to the right. There was very little danger that I'd notice her probable position for myself, because with sudden explosions like that—three shots in as many seconds, coming as a complete surprise and in circumstances where one can see nothing—it's extremely difficult to judge direction at all accurately. The only thing I knew for certain was that they were fired from outside, and perhaps from above ground.

"To go on: as soon as Lina had done her share in my deception it was up to her to get down from the tree as quickly as possible, put the revolver where Hinkson could find it, and then rush indoors to her own room to change: all of which she did. Meanwhile, the result of the shots upon myself was what had been so cleverly anticipated. I stepped forward a couple of paces, thus giving Hinkson his chance to prepare his escape.

"He was behind the door, as I've said: but not as himself. He was disguised as you, with eye-shade and tailcoat and wig. You hadn't followed me into the room at all, as I naturally supposed later, but had stood hovering in the doorway. The moment I moved, Hinkson did too, silently darting round the door, putting his right hand on the outer handle, and pretending to be you looking for a concealed murderer. Then, apparently in consequence of the shots, and for my benefit in case I turned round before he was ready, he clapped his left hand to his head. Finally, in order to attract my notice, he made a noise.

"I did turn then, and what I thought I saw was you being badly frightened. Bearing in mind that I'd been with you for only a few minutes, in a bad light, and that I'd never set eyes on Hinkson at all, I don't think I'm altogether to be blamed for falling into the trap. The only things I knew about your appearance were your height, your build, your dress, and a general impression of your features, including the all-important eye-shade: and all those particulars I either observed in Hinkson, or was given a reason for not observing. Thus, I saw your coat-tails flying as you ran out, your dress and height and eye-shade seemed just the same as before, and I didn't expect to see your face because you had your left hand to it—on the unshaded side—in a gesture of frightened horror.

"Well, having duly got me to look at him, Hinkson departed in a perfect imitation of a gun-shy man in a panic: and I remained firmly convinced that that man was yourself. I called you back, and you came: literally you, Hinkson by then having disappeared into one of the empty rooms on the other side of the Long Corridor. Thence, I imagine, he nipped through the window on to the drive, and tore off to wherever he'd hidden his dinner-jacket suit. And now, of course, I appreciate why the two rooms opposite the study were swept so clean: not for a dance, but so that there shouldn't be any tell-tale footmarks in the dust.

"If I'd come to the door to recall you, I still shouldn't have noticed anything amiss: you'd have been returning shamefacedly, with Hinkson well out of sight. The door through which he went was ready for him just ajar, I expect, and something soft wedged against the wall to prevent any noise when he pushed it open. Also, while I'm talking about noise, it's significant that Mr. Purdon kept mentioning your quietness of foot in his letters. You had to have that reputation, so that I shouldn't afterwards wonder why I hadn't heard you follow me in.

"There are three other points worth referring to in connection with what I've already dealt with. First, the eye-shades: Hinkson had naturally provided himself beforehand with two identical ones, and the excuse for discarding Mr. Purdon's highly-coloured purchase was excellent. The Major didn't like it, and had given you one of his own.

"Second, Hinkson had an explanation—after a suitable delay for apparent thought on the matter—for the fact that all the trees seemed to have been climbed. Adair had done that himself, it was alleged, and if I hadn't made the suggestion about looking at his clothes, all convincingly faked, it would have been made for me.

"Third comes the interesting point that as I stood in the study facing away from the door, there was nothing in front of me which could possibly reflect movements behind. The partition shut off the window, the panels were harmless, and the only picture was unglazed. As well, Hinkson had previously taken pains to establish from Mr. Purdon that I don't wear glasses.

"And now, having I hope made things clear so far, I'll go on to what happened next. Lina was safely upstairs getting into normal clothes—actually, the pink frock I saw her in later. Montague and Tilly and Mr. Purdon were also in their respective rooms, and all unable to hear shots fired from the little lawn: about that, too, Hinkson would have experimented on September 18th, in all probability. The cook had gone to bed, you were with me, Buck was presumably either in the woods or on his way to the house, and Hinkson himself was hurriedly changing somewhere out of doors. He'd already provided a suitable rig-out: dinner-jacket suit, shirt, and patent-leather shoes, all soiled as if from tree-climbing. The things he took off I expect he did up in a bundle together with the eye-shade, and hid until he could remove them next day.

"He was then ready to play his expected part: that of worried secretary in search of mysterious intruder. He collected his revolver from wherever Lina had put it, and nosed about the grounds until he encountered Buck. He could pretty well rely on that happening sooner or later; and he thought he could also rely on Buck's having been far enough away from the house not to hear the shot which killed Adair. My reasons for believing he was wrong about that I'll give presently.

"What I did next myself you know, and everything that took place later that evening, as far as your side of the story's concerned. You and Lina and Hinkson all acted as innocently as possible, and left me to it. At least, I'd better qualify that: Lina and Hinkson didn't act too well. Purposely they, and especially he, allowed me to feel that I hadn't heard quite the whole truth. They argued that by and by, when they came to me and made their confession, I should be more disposed to believe them because I'd already been dissatisfied with their first evidence. It's an ancient trick, but often effective. Meanwhile I could hardly miss the obvious solution of Adair's death, suicide: but in case I didn't fancy it for some personal and peculiar reason, there was always a pair of highly suspicious strangers to fall back on as possible suspects, either or both of them being the villainous Warner.

"And now I'm going to tackle things from the point of view of Hinkson's mistakes. I shouldn't be surprised to learn that he doesn't think he made any, and perhaps that's your opinion too, or was. All the same, he did—half a dozen or so.

"First, there was a bad discrepancy in the story told me on Saturday night. According to the girl, Hinkson lied to me about having gone to sleep in the beech tree through a desire to shield her theft of the ruby. Yet he didn't—possibly couldn't—contradict her when she later said that she didn't tell him she'd stolen the thing till three o'clock the next morning. I ought to have spotted that for a lie at the time, but somehow didn't.

"The next mistake was connected with their story of what they did on Thursday night. They said that while the others were playing bridge the two of them searched Tilly's room: Hinkson genuinely believing she might have stolen the stone, and Lina playing up to him. The reason why they didn't search Mr. Montague's room as well, he told me, was lack of time, and that was why they were forced to employ their acknowledged subterfuge the next evening, Friday, to get him out of the way. On reflection I really couldn't swallow that. If Hinkson had been as keen on recovering the ruby as he pretended, he'd have done one of the rooms on Thursday while Lina did the other, and thus I saw no real reason for Friday night's elaborate plotting. In view of my knowledge about the position of the various rooms in relation to reports from the study, it occurred to me that they might have wanted Montague out of earshot, and that alone.

"The third mistake was to do with the papers in the study. On Saturday afternoon Hinkson assured me in there that Adair couldn't have burnt all his notes in five minutes: yet later the same day Lina said in his presence that that was just what had happened. The Major had turned on her, told her that if she wanted his treasure she could look for it herself, and put his translation labours straight on the fire. Hinkson almost made another slip, too, in admitting to knowledge of the parchment's continued existence, but I muddled that myself. Like a fool I went to you for information, and you were able to explain that you'd overheard me talking to Mr. Purdon, and told Hinkson yourself.

"His fourth error—and perhaps the most careless of all—was in regard to Buck. Exactly what happened I can't prove—yet: but I can give you a pretty good idea. Buck *did* hear the shot which killed Adair, and came back towards the house to investigate. He then saw Lina climbing into the beech tree, and remained hidden but interested spectator of all that took place in the next half-hour. Afterwards, typically, he began to put his knowledge to good use by blackmailing the two of them, especially the girl. That was why she went off for walks with him, and explains the quarrel I almost interrupted on Sunday morning. It also explains why Buck and Hinkson so noticeably changed their previous footing of mutual dislike. Buck of course thought he was on to a good thing, and didn't want the suicide idea crabbed unless he did it himself. He knew practically the whole murder plot, combining what he saw with what Hinkson unsuspectingly told him when they met outside. Hinkson, on the other hand, just had to be pleasant to Buck—though he gave me a characteristically ingenious reason for his change of attitude.

"Buck, by the way, helped me considerably, though doubtless without meaning to. On Friday evening, when I first saw him in the dining-room, he hinted fairly broadly that there might have been something fishy about your black eye, and afterwards I remembered that.

"Hinkson's fifth mistake was over Adair's finished translation to the parchment. He couldn't make sense of it, and so hid the paper in the study, putting a clue to the hiding-place in Adair's pocket-diary. If I found it, and made sense of it, well and good: he'd back his wits to diddle everybody even then. If not, he'd wait till he could be sure it was safe, and then—providing he still couldn't puzzle out its meaning—devise some means for drawing my attention to the book he'd put it in. I thought all along how improbable it would have been for Adair himself to hide his own translation, especially in a place like that when he'd got a safe upstairs. Either he'd have destroyed it with the rest of the stuff, and the parchment too, by the way, or he'd have kept it on his person or in his desk.

"Hinkson's last mistake, though I say it myself, was in underrating my abilities. He never fully realized what a suspicious turn of mind I have, nor that I'd be capable of setting a trap for him by leaving that paper in the study apparently unnoticed. Some time between Tuesday evening and ten o'clock on Wednesday morning he couldn't restrain his curiosity any longer. He went to see if the paper had been removed, and by doing so practically convinced me that Adair's death was a case of murder. He was clever enough at not provoking suspicion—in refusing to work in the study, and in appearing to be enthusiastic about the idea of having a go at the parchment himself: but he couldn't conceive the possibility that I might have been on the watch for hocus right from the moment I saw the body. That's his main fault—over-confidence: vanity. He even had the cheek to put Adair's fishing-rod in Tilly's room, for the benefit of anybody silly enough to bother with impracticable conjuring tricks.

"As things stood, there was only the one way for a murder to have been done in that room: by the murderer's remaining inside until after my entry. Mr. Montague did suggest the idea, as a matter of fact, and so did Mr. Purdon again on Saturday night, but I ridiculed the idea that anyone could have sneaked out unseen. As it turned out I was right, too. I observed the murderer depart under my nose, and thought I was looking at a gun-shy man in a fright.

"In addition, once I understood how the deed was done I had no doubt about the identity of its combined authors. It really was you who called me out of Mr. Purdon's room, because you couldn't have risked his spotting a disguise and a different voice. From that it followed that it couldn't also have been you who went out of the study, since that person was the murderer, in there since firing the shot. Therefore it was Hinkson, the only other man in the house of your build. You'll doubtless remember that Mr. Purdon remarked about the similarity when he stood you a drink in Cheltenham.

"Again, the person who fired the shots outside to keep my attention away from Hinkson behind the door was yet another conspirator. It couldn't have been Tilly, whose eyesight is poor, and it wasn't likely to be Mr. Montague, since men of sixty-nine aren't usually much good at tree-climbing. Therefore it was either Judd or Buck or Lina. Judd I could account for elsewhere, and anyway I couldn't see Hinkson ever making an ally of him. Buck's story wasn't satisfactory, but I was certain that the planning had started last September, when Hinkson probably didn't know of his existence. That left Lina. Their open collaboration over the supposed loss of the ruby suggested that she was the third person in the affair, and the black knickers settled it as far as I was concerned.

"And while I'm talking about her I may as well point out a serious mistake she made. Whether Hinkson even now knows that she ever played about with Judd I can't say, but I very much doubt it. In any event, I'm quite sure he wouldn't have approved her behaviour with Montague in her room up to quarter to ten on Friday night. To let him kiss her 'good and hard', which is what Judd assures me happened, isn't at all consistent with her story: she was supposed to be extremely upset, and longing to get rid of the man the minute it was safe to.

"Well, Andrews, what do you think of my tale?"

The butler had long since recovered his wits. His interest was unmistakable, though, and not always as deferential as I think he meant it to be. As for the constable, Johnson, he remained stolidly against the wall, and only his eyes showed that he was awake and alert.

"Frankly, sir, it astounds me," said Andrews now. "I hardly know whether you wish me to take it seriously or not."

"Then you'd better be quite clear on the point," I told him: "I do. Now, why do you suppose I came to you tonight, instead of going to Lina or Hinkson?"

"Well sir, I can't imagine—I really don't know."

"Not? I think you mean that you won't admit you know. You see, I'm absolutely certain that the account of the murder I've just given you is true—as certain as that I'm sitting here on the end of your bed. I'm equally sure that you're well aware I'm right, too, and what I particularly wanted to do was to show you just how much I do know. You see, whether you choose to admit it or not, in your heart you're now a badly frightened man, and from my point of view that's just as well, because I'm going to frighten you even more in a moment."

Andrews smiled slightly, and I had to admire his coolness.

"Yes sir?" he murmured: as if I had just threatened to relate some not very interesting information at great length.

"Yes," I said, and nodded as seriously as possible. It was vital for me to impress him, and I fancied from a slight change of expression on his shadowed face that I might be succeeding.

"What I want you to do is to consider your position," I went on. "In my opinion you're Hinkson's accomplice, and I've explained why I think that. Now, perhaps you're bracing yourself with the belief that he'll stick by you as long as you stick by him. But if so, you're being extraordinarily foolish, and I'll try to show you why. Hinkson planned and executed Adair's murder for only one reason: to get hold of the treasure. What made him believe in it heaven knows, but he did, and more so when the tangible ruby turned up. Well, as far as he's concerned the necessity for the treasure no longer exists. Do you understand me?"

"No sir, I'm afraid not."

"Ah, I thought you mightn't—I thought perhaps he wouldn't tell you that part. The fact is, Andrews, I found Major Adair's will in his safe, and I'll let you read it if you promise not to snatch."

Putting the light in a better position, I allowed him to digest the contents of the document, and felt fairly sure that they came as a surprise.

"And they tell me the estate will be worth well over a hundred thousand pounds," I added. "Now do you understand?"

"No sir—not altogether."

"But you're beginning to, I think? Look at things from Hinkson's point of view if you can, and remember that he hasn't the remotest idea I suspect anything. No, don't look so relieved: it's nothing for you to be glad about. This is what he's working out in that sly subtle mind of his.

" 'Here am I a murderer because of a treasure, and I'm not a step nearer laying hands on it than I was last year. But I *am* nearer a fortune: I've only got to marry Lina, and there we are. Except for Andrews and Buck: *they know too much.*' "

I paused there, significantly, and saw the butler's fleeting glance of apprehension.

"Yes, that's what he's thinking" I said. "If he marries Lina he's safe from her, because in law a wife can't be compelled to give evidence against her husband. He's also getting close to money, big money, and that was what he wanted so badly that he didn't mind killing Adair for it. But he isn't safe from Buck, and he isn't safe from you, and he doesn't stand to gain a penny from your continued existence. So what does he do?"

Andrews remained silent, still playing his part: but with less conviction.

"Well, man, isn't it obvious? He goes and buys a copy-book, and among the old maxims about stitches in time and many hands, and best policies he finds one which appeals to him more than all the rest put together: *Dead men tell no tales.*

"Yes, you may well squirm a bit at that: and Buck would have done too—only he's out of it."

"Why?" asked Andrews sharply: the first involuntary proof that he was concerned.

"Because Buck happens to be an American kidnapper," I explained. "He's urgently wanted across the Atlantic, and so I arrested him at ten o'clock tonight, and by now he's safe in gaol. You see, it's my job not only to, come along after murders have been done, and clear them up, but also if I can to prevent them from happening."

That really shook him, I saw, and hastily I pressed home my advantage.

"And Hinkson isn't going to like it at all when he gets to hear," I said. "His first thought will be 'How long for Buck to spill what he knows? How long have I got to marry the girl and get clear?'. And his second thought will be that getting clear won't do him an atom of good unless he silences Buck first: but how?

"Again, the answer's obvious: by a bribe big enough to make Buck cheerfully endure a few years of prison in silence for the sake of what he'll have to spend when he gets out. For instance, say twenty thousand pounds in cash, delivered to Buck before he's extradited. But Hinkson *can't* deliver it till he's married Lina; and he can't marry Lina unless he's at liberty; and he knows very well his liberty isn't safe while you're alive.

"So, I repeat, what does he do?"

(As has doubtless been noticed, the argument in the last paragraph isn't particularly strong, but I wasn't worrying. By now Andrews looked far too agitated to think logically. His brow was agleam again, his hands restless, and his one exposed eye blinking nervously.)

"Well," I said, as he remained silently perturbed, "once more the answer's plain. He employs every ounce of cunning he possesses to get rid of you as soon as possible: and you know as well as I do that cunning is Hinkson's strong suit. Just lie there and think about it for a moment: it may be advisable."

Now, I had only my own judgment to rely on about whether or not to arrest Andrews. Because of the stringent rules about such matters I could neither threaten him into a confession nor offer him any material inducement to make one. Up till now I'd been justified in all I'd said, because legally silence can be as much an admission of agreement as speech. All the same it was a delicate position, and I felt glad that I had Johnson as a witness.

It can be imagined that during the whole time I was speaking I had watched Andrews with extreme attention. I had no doubt in my own mind that my reconstruction of the crime was accurate: the trouble was that I couldn't see how to obtain proof. My one hope seemed to lie in making the butler believe, without suggesting it in so many words, that the best thing he could do for himself was to tell everything he knew. He would have to do this voluntarily, though, and must first be charged with some specific offence.

There isn't much more to relate. For perhaps three minutes I scanned his face anxiously, estimating my chances, and everything I saw led me to the conclusion that I could chance my arm. I chose my moment either well or luckily, arrested him for complicity in the murder of Adair, and administered the usual caution: and to my inestimable relief he broke down.

If anyone thinks I could have tackled the problem in any better way, I shall be glad to have details. For myself, I could see no other, and in any case I believe that Andrews did the wisest thing. That Hinkson *would* eventually have murdered him, given the proverbial half a chance, I am convinced.

It took Johnson and myself till five o'clock before we had his full statement down, read through, and duly signed and witnessed, and after that the arrest of Hinkson and Lina was child's play. Contrary to my expectation, the secretary took things with comparative calmness: he turned deathly pale, but didn't collapse or start chattering. The girl behaved very differently, though. I should never have believed that so seductive a nightdress could have contained so vigorous a fiend, and I am sure she would have killed us both if she could have got her hands on a gun.

They're both dead themselves now, so I mustn't say too much, but at the time I could scarcely raise a whisper of pity in my heart for either. In Hinkson's room we found valuable evidence: his eye-shade, the clothes he had worn when impersonating Andrews, and about fifteen grains of *Pernocton,* the hypnotic the police-surgeon had told me about on the telephone. "We found something else, too, which confirmed a suspicion of mine that I haven't yet mentioned: a second revolver, fully loaded and with a silencer attached. I can't help thinking that had anything gone astray with his plans after I entered the study, that would have been the end of me.

As for Andrews, he also is gone, but not by the rope. The doctors said it was *angina pectoris,* and I feel that he just deserved the privacy of a hospital to die in.

APPENDIX

(The solution of the clue to Jasper Mauberley's treasure.)

And now for the gold and jewels that I have hid, & how a man must act who has hope in his covetous heart of obtaining them. Yet gold alone will not bring him to heaven, therefore do I call him covetous. Happy is the man that findeth wisdom, & the man that getteth understanding; for the merchandise of it is better than the merchandise of silver, & the gain thereof than fine gold (1). Wisdom is the principal thing; therefore get wisdom, & with all thy getting get understanding (2). Ponder the path of thy feet, & let all thy ways be established (3).

Let him take one pace to the north, & four to the east, & three to the south, & four to the west; & let him not be troubled that the measurements are inaccurate & the directions of no account. And for a sign, let him be polluted to the waist with blood.

The first thing that struck me was the biblical turn of language employed. I was almost sure I remembered having read the sentence I have marked (1) before, and some of the rest of it seemed vaguely familiar. As I have said in Chapter XV, the large book-case contained among other things a fairly complete reference library. Included in this was a copy of Cruden's *Concordance*—'A Dictionary and Alphabetical Index to the Bible', as it is described on the title-page. By means of it I discovered that the translated clue contained three quotations, exact but for two slight alterations: the running together of two sentences in (1), and a comma for a colon in (2).

These "quotations are
>   (1) : Proverbs III, 13, 14;
>   (2) : Proverbs IV, 7;
>   (3): Proverbs IV, 26.

The fact that they seemed so apposite to the subject, and fitted so neatly into place with regard to wording, made me think that perhaps Jasper Mauberley intended the Old Testament to spring to the reader's mind. (*Conclusion 1.*)

For some time that was as far as I got: I could make nothing of the second paragraph. Then it occurred to me that the word *directions* might mean *lines of course* and not *instructions:* an interpretation I was the more willing to accept after drawing the following diagram several times:

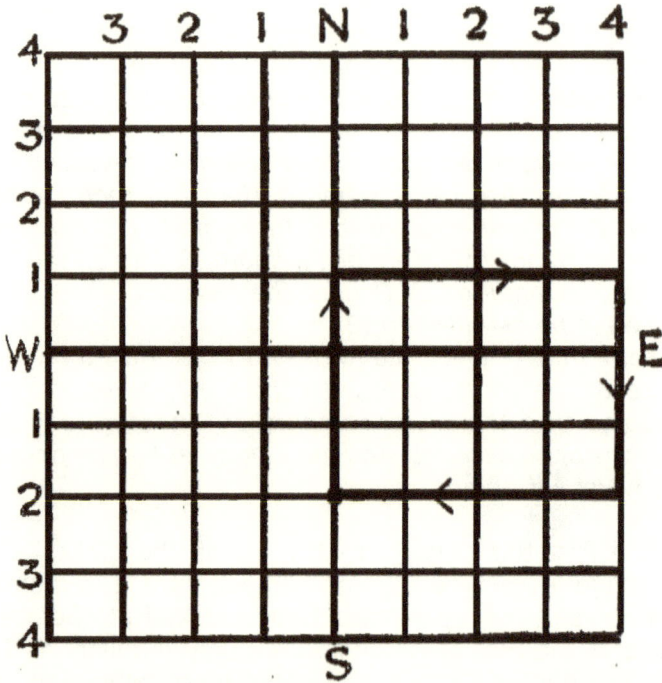

But ready as I was to dispense with the cardinal points of the compass, I didn't see how I could afford to do the same with the specific numbers of paces: otherwise I should have nothing definite left. Yet the second paragraph said plainly that the measurements were inaccurate. I finally evolved two excuses for getting over the difficulty with a clear conscience.

(a) The word *measurements* might not refer to the paces mentioned.

(b) It might be that it was the word *pace* which was wrongly used, perhaps for *mile;* but the prefixed numbers—or perhaps only their mutual ratios—remained as stated.

In any event, I decided to retain the numbers, but to discard the points of the compass. *(Conclusion 2.)*

After further thought it crossed my mind that possibly the numbers had nothing to do with distances. After that it was no great step to suppose that *Conclusions 1* and *2* could be related: the numbers referred to books or chapters or verses in the Bible.

Before going on I devoted some moments to a consideration of whether this was at all probable, and decided in favour. I could think of scarcely any other book which would so aptly fulfill what must have been the three essential conditions for employment as a hiding-place for a clue.

First, the book must be extant in 1692.

Second, it must be likely to be known to Mauberley.

Third, it must be such that he could have every expectation of its still being extant and widely known for very many years to come.

Very well, I said to myself: the numbers mentioned—1, 4, 3, 4—conceal something in the Bible; almost certainly a verse. At a guess, too, it will be a verse in the Old Testament, since all the quotations in the first paragraph came from there.

I worked out eleven possibilities:—

(1) : the 1434th verse;
(2) : the 143rd and 144th verses;
(3) : the 143rd chapter, the 4th verse;
(4) : the 14th chapter, the 34th verse;
(5) : the 14th chapter, the 3rd and 4th verses;
(6) : the 1st book, the 43rd chapter, the 4th verse;
(7) : the 1st book, the 4th chapter, the 34th verse;
(8) : the 1st book, the 4th chapter, the 3rd and 4th verses;
(9) : the 14th book, the 3rd chapter, the 4th verse;
(10): the 14th book, the 3rd and 4th verses;
(11): the 143rd book, the 4th verse.

I began methodically searching for the above references, and the very first discouraged me. It commenced *And there was no bread in all the land. . .* The second proved equally disheartening. I then recalled the sign promised in the last line or so, and resorted again to Cruden. Under *polluted* I found four references to blood:

(1) : Psalm CVI, 38;
(2) : Lamentations IV, 14 5
(3) : Ezekiel XVI, 6;
(4) : Hosea VI, 8.

Translating into terms of books, chapters, and verses, these became:

(1): 19:106: 38;
(2): 25: 4:14;
(3): 26:16: 6;
(4): 28: 6: 8.

This seemed of little help until it occurred to me (I'm sorry to keep repeating the phrase, but I don't know how else to express what happened) that *to the waist* was a rough synonym for *half*. In other words, the sign would be twice the books and/or chapters and/or verses in the quotation which was the interpretation of the clue.

After that, of course, it was easy. Only (3) and (4) halved decently, and 13: 8: 3 wasn't among my eleven possibilities. 14: 3: 4 was, though—the ninth on the list.

The sign was Hosea VI, 8, and the clue would be II Chronicles III, 4; and a glance at II Chronicles settled the question beyond doubt.

And the porch that was in the front of the house, the length of it was according to the breadth of the house, twenty cubits, and the height was an hundred and twenty: and he overlaid it within with pure gold.

It will be remembered—or can be verified—that Tony mentioned the porch near the beginning of Chapter V, and I did so myself in describing my arrival at the house. It will also be understood that the measurements were indeed likely to be inaccurate, because a hundred and twenty cubits is at least sixty yards.

Some time after the events narrated in this book a proper search was made, and a considerable quantity of gold and precious stones found. Adair would have been disappointed, though: the total value did not exceed £15,000. Most of this treasure went to Tilly, and for my assistance she proposed that I take half. That was absurd, of course, and in the end I satisfied her by accepting four of the smaller gems, an emerald, a ruby, a diamond, and a sapphire; so if one day you see a man wearing odd and obviously valuable cufflinks, it may—but probably won't—be me.

THE END

RAMBLE HOUSE's

HARRY STEPHEN KEELER WEBWORK MYSTERIES

(RH) indicates the title is available ONLY in the RAMBLE HOUSE edition

The Ace of Spades Murder
The Affair of the Bottled Deuce (RH)
The Amazing Web
The Barking Clock
Behind That Mask
The Book with the Orange Leaves
The Bottle with the Green Wax Seal
The Box from Japan
The Case of the Canny Killer
The Case of the Crazy Corpse (RH)
The Case of the Flying Hands (RH)
The Case of the Ivory Arrow
The Case of the Jeweled Ragpicker
The Case of the Lavender Gripsack
The Case of the Mysterious Moll
The Case of the 16 Beans
The Case of the Transparent Nude (RH)
The Case of the Transposed Legs
The Case of the Two-Headed Idiot (RH)
The Case of the Two Strange Ladies
The Circus Stealers (RH)
Cleopatra's Tears
A Copy of Beowulf (RH)
The Crimson Cube (RH)
The Face of the Man From Saturn
Find the Clock
The Five Silver Buddhas
The 4th King
The Gallows Waits, My Lord! (RH)
The Green Jade Hand
Finger! Finger!
Hangman's Nights (RH)
I, Chameleon (RH)
I Killed Lincoln at 10:13! (RH)
The Iron Ring
The Man Who Changed His Skin (RH)
The Man with the Crimson Box
The Man with the Magic Eardrums
The Man with the Wooden Spectacles
The Marceau Case
The Matilda Hunter Murder
The Monocled Monster
The Murder of London Lew
The Murdered Mathematician
The Mysterious Card (RH)
The Mysterious Ivory Ball of Wong Shing Li (RH)
The Mystery of the Fiddling Cracksman
The Peacock Fan

The Photo of Lady X (RH)
The Portrait of Jirjohn Cobb
Report on Vanessa Hewstone (RH)
Riddle of the Travelling Skull
Riddle of the Wooden Parrakeet (RH)
The Scarlet Mummy (RH)
The Search for X-Y-Z
The Sharkskin Book
Sing Sing Nights
The Six From Nowhere (RH)
The Skull of the Waltzing Clown
The Spectacles of Mr. Cagliostro
Stand By—London Calling!
The Steeltown Strangler
The Stolen Gravestone (RH)
Strange Journey (RH)
The Strange Will
The Straw Hat Murders (RH)
The Street of 1000 Eyes (RH)
Thieves' Nights
Three Novellos (RH)
The Tiger Snake
The Trap (RH)
Vagabond Nights (Defrauded Yeggman)
Vagabond Nights 2 (10 Hours)
The Vanishing Gold Truck
The Voice of the Seven Sparrows
The Washington Square Enigma
When Thief Meets Thief
The White Circle (RH)
The Wonderful Scheme of Mr. Christopher Thorne
X. Jones—of Scotland Yard
Y. Cheung, Business Detective

Keeler Related Works

A To Izzard: A Harry Stephen Keeler Companion by Fender Tucker — Articles and stories about Harry, by Harry, and in his style. Included is a compleat Keeler bibliography.

Wild About Harry: Reviews of Keeler Novels — Edited by Richard Polt & Fender Tucker — 22 reviews of works by Harry Stephen Keeler from *Keeler News*. A perfect introduction to the author.

The Keeler Keyhole Collection: Annotated newsletter rants from Harry Stephen Keeler, edited by Francis M. Nevins.

Fakealoo — Pastiches of the style of Harry Stephen Keeler by selected demented members of the HSK Society.

RAMBLE HOUSE
Fender Tucker, Prop.
www.ramblehouse.com   fender@ramblehouse.com
318-455-6847   443 Gladstone Blvd. Shreveport LA 71104

RAMBLE HOUSE's OTHER LOONS

Freaks and Fantasies — Eerie tales by Tod Robbins, collaborator of Tod Browning on the film FREAKS.

Vixen Scandal — Two sleaze masterpieces from the 60s by Jim Harmon: *Vixen Hollow* and *Celluloid Scandal.*

The Koky Sundays — A collection of all of the 1978-1981 Sunday comic strips by Richard O'Brien and Mort Gerberg.

Marblehead: A Novel of H.P. Lovecraft — A long-lost masterpiece from Richard A. Lupoff. Published for the first time!

The Secret Adventures of Sherlock Holmes — Three Sherlockian pastiches by the Brooklyn author/publisher, Gary Lovisi.

The Universal Holmes — Richard A. Lupoff's 2007 collection of five Holmesian pastiches.

The Singular Problem of the Stygian House-Boat — Two classic tales by John Kendrick Bangs about the denizens of Hades.

The One After Snelling — Kickass modern noir from Richard O'Brien.

Tales of the Macabre and Ordinary — Modern twisted horror by Chris Mikul, author of the *Bizarrism* series.

The Gold Star Line — A classic seaboard adventure from L.T. Reade and Robert Eustace.

The Werewolf vs the Vampire Woman — Hard to believe ultraviolence by either Arthur M. Scarm or Arthur M. Scram.

Black Hogan Strikes Again — From Australia's Peter Renwick comes a tale of the outback.

Four Joel Townsley Rogers Novels — By the author of *The Red Right Hand: Once In a Red Moon, Lady With the Dice, The Stopped Clock, Never Leave My Bed*

Night of Horror — A short story collection of Joel Townsley Rogers

Twenty Norman Berrow Novels — *The Bishop's Sword, Ghost House, Don't Go Out After Dark, Claws of the Cougar, The Smokers of Hashish, The Secret Dancer, Don't Jump Mr. Boland!, The Footprints of Satan, Fingers for Ransom, The Three Tiers of Fantasy, The Spaniard's Thumb, The Eleventh Plague, Words Have Wings, One Thrilling Night, The Lady's in Danger, It Howls at Night, The Terror in the Fog, Oil Under the Window, Murder in the Melody, The Singing Room*

The N. R. De Mexico Novels — Robert Bragg presents *Marijuana Girl, Madman on a Drum, Private Chauffeur* in one volume.

Two Hake Talbot Novels — *Rim of the Pit, The Hangman's Handyman.* Classic locked room mysteries.

Two Alexander Laing Novels — *The Motives of Nicholas Holtz* and *Dr. Scarlett*, stories of medical mayhem and intrigue from the 30s.

Two Wade Wright Novels (and counting) — *Echo of Fear* and *Death At Nostalgia Street*, with more to come!

Five Jack Mann Novels — Strange murder in the English countryside. *Gees' First Case, Nightmare Farm, Grey Shapes, The Ninth Life, The Glass Too Many.*

Three Max Afford Novels — *Owl of Darkness, Death's Mannikins* and *Blood on His Hands* by One of Australia's finest novelists.

Five Joseph Shallit Novels — *The Case of the Billion Dollar Body, Lady Don't Die on My Doorstep, Kiss the Killer, Yell Bloody Murder, Take Your Last Look.* One of America's best 50's authors.

The Anthony Boucher Chronicles — edited by Francis M. Nevins
Book reviews by Anthony Boucher written for the *San Francisco Chronicle*, 1942 – 1947. Essential and fascinating reading.

The Best of 10-Story Book — edited by Chris Mikul, over 35 stories from the literary magazine Harry Stephen Keeler edited.

A Young Man's Heart — A forgotten early classic by Cornell Woolrich

Muddled Mind: Complete Works of Ed Wood, Jr. — David Hayes and Hayden Davis deconstruct the life and works of a mad genius.

My First Time: The One Experience You Never Forget — Michael Birchwood — 64 true first-person narratives of how they lost it.

The Incredible Adventures of Rowland Hern — Rousing 1928 impossible crimes by Nicholas Olde.

Don Diablo: Book of a Lost Film — Two-volume treatment of a western by Paul Landres, with diagrams. Intro by Francis M. Nevins.

The Charlie Chaplin Mystery — Movie hijinks by Wes D. Gehring.

Gamefinger — 1966 sado-sleaze from William Knoles.

Dime Novels: Ramble House's 10-Cent Books — *Knife in the Dark* by Robert Leslie Bellem, *Hot Lead* and *Song of Death* by Ed Earl Repp, *A Hashish House in New York* by H.H. Kane, and five more.

Stakeout on Millennium Drive — Indianapolis Noir — Ian Woollen.

Dope Tales #1 — Two dope-riddled classics; *Dope Runners* by Gerald Grantham and *Death Takes the Joystick* by Phillip Condé.

Dope Tales #2 — Two more narco-classics; *The Invisible Hand* by Rex Dark and *The Smokers of Hashish* by Norman Berrow.

Dope Tales #3 — Two enchanting novels of opium by the master, Sax Rohmer. *Dope* and *The Yellow Claw.*

Tenebrae — Ernest G. Henham's 1898 horror tale brought back.

The Sign of the Scorpion — 1935 Edmund Snell tale of oriental evil.

The House of the Vampire — 1907 thriller by George S. Viereck.

An Angel in the Street — Modern hardboiled noir by Peter Genovese.

The Devil's Mistress — Scottish gothic tale by J. W. Brodie-Innes.

The Lord of Terror — 1925 mystery with master-criminal, Fantômas.

The Lady of the Terraces — 1925 adventure by E. Charles Vivian.

My Deadly Angel — 1955 Cold War drama by John Chelton.

Prose Bowl — Futuristic satire — Bill Pronzini & Barry N. Malzberg .

Satan's Den Exposed — True crime in TorC New Mexico — Award-winning journalism by the Desert Journal.

The Amorous Intrigues & Adventures of Aaron Burr — by Anonymous — Hot historical action.

I Stole $16,000,000 — True story by cracksman Herbert E. Wilson.

The Black Dark Murders — Vintage 50s college murder yarn by Milt Ozaki, writing as Robert O. Saber.

Sex Slave — Potboiler of lust in the days of Cleopatra — Dion Leclerq.

You'll Die Laughing — Bruce Elliott's 1945 novel of murder at a practical joker's English countryside manor.

The Private Journal & Diary of John H. Surratt — The memoirs of the man who conspired to assassinate President Lincoln.

Dead Man Talks Too Much — Hollywood boozer by Weed Dickenson

Red Light — History of legal prostitution in Shreveport Louisiana by Eric Brock. Includes wonderful photos of the houses and the ladies.

Gadsby — A lipogram (a novel without the letter E). Ernest Vincent Wright's last work, published in 1939 right before his death.

A Snark Selection — Lewis Carroll's *The Hunting of the Snark* with two Snarkian chapters by Harry Stephen Keeler — Illustrated by Gavin L. O'Keefe.

Ripped from the Headlines! — The Jack the Ripper story as told in the newspaper articles in the *New York* and *London Times.*

Geronimo — S. M. Barrett's 1905 autobiography of a noble American.

The Compleat Calhoon — All of Fender Tucker's works: Includes *The Totah Trilogy, Weed, Women and Song* and *Tales from the Tower,* plus a CD of all of his songs.